Lecture Notes in Artificial Intelligence   4826

Edited by J. G. Carbonell and J. Siekmann

Subseries of Lecture Notes in Computer Science

Petra Perner Ovidio Salvetti (Eds.)

Advances in Mass Data Analysis of Signals and Images in Medicine Biotechnology and Chemistry

International Conferences, MDA 2006/2007
Leipzig, Germany, July 18, 2007
Selected Papers

 Springer

Series Editors

Jaime G. Carbonell, Carnegie Mellon University, Pittsburgh, PA, USA
Jörg Siekmann, University of Saarland, Saarbrücken, Germany

Volume Editors

Petra Perner
IBAI - Institute of Computer Vision and Applied Computer Science
Arno-Nitzsche-Str. 43, 04277 Leipzig, Germany
E-mail: pperner@ibai-institut.de

Ovidio Salvetti
Italian National Research Council (CNR
Institute of Information Science and Technologies (ISTI)
via G. Moruzzi 1, 56124 PISA, Italy
E-mail: Ovidio.Salvetti@isti.cnr.it

Library of Congress Control Number: 2007940157

CR Subject Classification (1998): H.2.8, I.4.6, J.3

LNCS Sublibrary: SL 7 – Artificial Intelligence

ISSN   0302-9743
ISBN-10   3-540-76299-X Springer Berlin Heidelberg New York
ISBN-13   978-3-540-76299-7 Springer Berlin Heidelberg New York

Springer is a part of Springer Science+Business Media

springer.com

© Springer-Verlag Berlin Heidelberg 2007
Printed in Germany

Typesetting: Camera-ready by author, data conversion by Scientific Publishing Services, Chennai, India
Printed on acid-free paper   SPIN: 12181568   06/3180   5 4 3 2 1 0

Preface

The automatic analysis of images and signals in medicine, biotechnology, and chemistry is a challenging and demanding field.

Signal-producing procedures by microscopes, spectrometers, and other sensors have found their way into wide fields of medicine, biotechnology, economy, and environmental analysis. With this arises the problem of the automatic mass analysis of signal information. Signal-interpreting systems which generate automatically the desired target statements from the signals are therefore of compelling necessity. The continuation of mass analyses on the basis of classical procedures leads to investments of proportions that are not feasible. New procedures and system architectures are therefore required.

The scope of the International Conference on Mass Data Analysis of Images and Signals in Medicine, Biotechnology and Chemistry MDA (www.mda-signals.de) is to bring together researchers, practitioners, and industry people who are dealing with mass analysis of images and signals to present and discuss recent research in these fields.

The goals of this workshop are to:

- Provide a forum for identifying important contributions and opportunities for research on mass data analysis on microscopic images
- Promote the systematic study of how to apply automatic image analysis and interpretation procedures to that field
- Show case applications of mass data analysis in biology, medicine, and chemistry

Topics of interest include (but are not limited to):

- Techniques and developments of signal and image producing procedures
- Object matching and object tracking in microscopic and video microscopic images
- 1D, 2D, and 3D shape analysis and description
- 1D, 2D, and 3D feature extraction of texture, structure, and location
- Algorithms for 1D, 2D, and 3D signal analysis and interpretation
- Image segmentation algorithms
- Parallelization of image analysis and interpretation algorithms
- Semantic tagging of images from life science applications
- Applications in medicine, biotechnology, chemistry, and others
- Applications in crystallography
- Applications in proteomics
- Applications in 2D and 3D cell images analysis
- Image acquisition procedures for mass data analysis

This volume is a post-proceedings of papers from MDA 2006 and MDA 2007. A large number of the papers propose new image-segmentation techniques for biological and medical applications. Image segmentation is a crucial step in image processing and the accuracy of this step heavily influences the final result. In the methodology the authors use they try to identify classes of images first and then they propose algorithms that should work robustly and accurate enough for this class of images.

The second portion of papers deals with new applications where imaging and signal-interpretation methods are used. These imaging methods range from optical methods to ultra-sonic microscopy. The applications are in air monitoring for hazardous materials, quality control of cereals, proteomics and drug design, as well as in the characterization of piezo-electric properties. Spectrometers are used for algae classification.

Other papers deal with specific topics such as semantic tagging of biological images, shape characterization under time-varying conditions, and statistical analysis of time-series for DNA sequencing.

Altogether, we were pleased to see how many different problems for imaging and signal interpretation have been presented, showing that there is a tremendous need for automatic methods. We hope we have managed to bring these problems into the center of attention and inspire many other researchers to work on these real applications.

The next International Conferences on Mass Data Analysis of Signals and Images will be held in July 2008. We are looking forward to your submissions.

July 2007   Petra Perner
                                                                            Ovidio Salvetti

International Conference on Mass Data Analysis of Images and Signals in Medicine, Biotechnology and Chemistry

MDA 2007 / 2006

Co-chairs

Petra Perner IBaI, Germany
Ovidio Salvetti CNR-ISTI, Italy

Program Committee

Walter Arnold Fraunhofer IzfP, Germany
Ewert Bengsston University of Uppsala, Sweden
Hans du Buf University of Algarve, Portugal
Eugenio Fava MPI-CBG, Germany
Maria Frucci Istituto di Cibernetica, C.N.R., Italy
Igor Gurevich Academy of Science, Russia
Thomas Günther JenaBios GmbH, Germany
Giulio Iannello Campus Bio-Medico of Rome, Italy
Xiaoyi Jiang University of Muenster, Germany
Montse Pardas Politècnica de Catalunya, Spain
Thang Viet Pham OncoProteomics Laboratory, Netherlands
Gabriella Saniti di Baja Istituto di Cibernetica , Italy
Arnold Smeulders University of Amsterdam, Netherlands
Tuan Pham James Cook University, Australia
Julie Wilson York Structural Biology Laboratory, UK

Table of Contents

Image Acquisition and Analysis of Hazardous Biological Material in Air

Christoph Sklarczyk[1], Horst Perner[2], Hans Rieder[1], Walter Arnold[1],
and Petra Perner[2]

[1] Fraunhofer-Institute for Non-Destructive Testing (IZFP),
Bldg. E 3.1 University, D-66123 Saarbrücken, Germany
[2] Institute of Computer Vision and Applied Computer Sciences,
IBaI, Leipzig, Germany

Abstract. Human beings are exposed every day to bio-aerosols in the various fields of their personal and/or professional daily life. The European Commission has rules protecting employees in the workplace from biological hazards. Airborne fungi can be detected and identified by an image-acquisition and interpretation system. In this paper we present recent results on the development of an automated image acquisition, probe handling and image-interpretation system for airborne fungi identification. We explain the application domain and describe the development issues. The development strategy and the architecture of the system are described and some results are presented.

Keywords: Microscopic image acquisition, microbiological probe handling, image analysis, image interpretation, case-based object recognition, case-based reasoning.

1 Introduction

Airborne microorganisms are ubiquitously present in various indoor and outdoor environments. The potential implication of fungal contaminants in bio-aerosols on occupational health is recognized as a problem in several working environments. There is a concern on the exposure of workers to bio-aerosols especially in composting facilities, in agriculture, and in municipal waste treatment. The European Commission has therefore guiding rules protecting employees in the workplace from airborne biological hazards. In fact, there are an increasing number of incidents of building-related sickness, especially in offices and residential buildings. Some of these problems are attributed to biological agents, especially in relation to airborne fungal spores. However, the knowledge of health effects of indoor fungal contaminants is still limited. One of the reasons for this limitation is that appropriate methods for rapid and long-time monitoring of airborne microorganisms are not available.

Besides the detection of parameters relevant to occupational and public health, in many controlled environments the number of airborne microorganisms has to be kept below the permissible or recommended values, e.g. in clean rooms, in operating

P. Perner and O. Salvetti (Eds.): MDA 2006/2007, LNAI 4826, pp. 1–14, 2007.

theaters, and in domains of the food and pharmaceutical industry. Consequently, the continuous monitoring of airborne biological agents is a necessity for the detection of risks of human health as well as for the flawless operation of technological processes.

At present a variety of methods are used for the detection of fungal spores. The culture-based methods depend on the growth of spores on an agar plate and on the counting of colony-forming units [14]. Culture-independent methods are based on the enumeration of spores under a microscope, the use of a polymerase chain reaction or on DNA hybridization for the detection of fungi [14]. However, all these methods are limited by time-consuming procedures of sample preparation in the laboratory. This paper describes the development and the realization of an automated image-acquisition and probe handling unit of biologically dangerous substances and the automated analysis and interpretation of microscope images of these substances.

In the system described here, contaminated air containing bio-aerosols is collected in a defined volume via a carrier agent. They are recorded by an image-acquisition unit, counted, and classified. Their nature is determined by means of an automated image-analysis and interpretation system. Air samples are automatically acquired, prepared and transferred by a multi-axis servo-system to an image-acquisition unit based on a standard optical microscope with a digital color camera. This part of the system is described in Section 2. To obtain a sufficient image quality, special requirements have to be fulfilled by the image-acquisition unit which will be described in Section 3.

The variability of the biological objects is very broad. Given the constraints of the image acquisition, this variability is found in the appearance of the objects as well. There are no general features allowing one to discern the type of the detected fungi. In the system employed here, images are stored, and a more generalized description for the different appearances of the same objects is used. We will describe this novel case-based reasoning approach for the image analysis and its interpretation in Section 4. Finally, we summarize our work in Section 5.

2 System Requirements

The system to be developed should allow to collect dust and biological aerosols in well-defined volumes over microscope slides, deposit them there, image them with an appropriate method and count and classify them with an automated image analysis and interpretation method, in order to determine the following parameters from the images:

- Total number of airborne particles
- Classification of all particles according to the acquired image features
- Classification of biological particles, e.g. spores, fragments of fungal mycelia, and fragments of insects
- Number of respirable particles
- Total number of airborne particles of biological origin
- Number of dead particles of biological origin
- Number of viable and augmentable particles of biological origin

- Identification of species or geni exploiting the characteristic shapes of spores and pollen
- Proportion of airborne abiotic and biotic particles
- Proportion of dead and viable airborne microorganisms.

At the beginning of the project the following requirements concerning the optical and the mechanical system were defined:

- Color images should be produced in order to facilitate the separation of dead and living objects.
- It should be possible to generate images in at least three defined depths of field.
- A marker liquid like lactophenol should be used to further enhance the separation of dead and living objects (blue color for living objects). For that a cover slip is necessary in order to uniformly distribute the marker drop on the object slide.
- The object slide should be covered with an adhesive in order to fix the airborne germs.

Table 1. Strains of fungi used and selected properties of spores

Species	Strain no.	Spore shape	Spore color	Spore size [µm]
Alternaria alternata	J 37 (A[1])	Septated, clavate to ellipsoidal	Pale brown	$18 - 83 \times 7$-18
Aspergillus niger	i400 (B[2])	Spherical, ornamented with warts and spines	Brown	Ø 3.5 - 5
Rhizopus stolonifer	J 07 (A)	Irregular in shape, often ovoid to elliptical, striate	Pale brown	7-15 × 6-8
Scopulariopsis brevicaulis	J26 (A)	Spherical to ovoid	Rose-brown	5-8 × 5-7
Ulocladium botrytis	i171(B)	Septated, ellipsoidal	Olive-brown	18-38 × 11-20
Wallemia sebi	J 35 (A)	Cubic to globose	Pale-brown	Ø 2.5 – 3.5

[1](A): from culture collection of JenaBios GmbH, Jena, Germany.
[2](B): from the fungal stock collection of the Institute of Microbiology, University of Jena, Jena, Germany.

Six fungal strains representing species with different spore types were identified as important species in different environments (Tab. 1) by our industrial project partner JenaBios GmbH. A database of images from the spores of these species was produced and was the basis of our development. The number of imaged spore per species was about 30-50. Since no commercial system was known fulfilling all requirements, a corresponding system was developed which is described in what follows.

3 The Automated Imaging System

3.1 The Microscopic Image-Acquisition System

Following the specifications given in Section 2 we developed an automated probe-handling and digital image-acquisition system for taking microbiological material from air samples [12]. An existing optical Leitz microscope was upgraded and expanded in its hardware. A lens from Olympus with a magnification of 60X and a numerical aperture of 0.7 was used. Its focal length of 1.7 mm provided sufficient clearance between the lens and the object slide including the cover glass to avoid collisions due to their variability in thickness. The lens was inserted in an autofocusing device from Physik Instrumente (PI, Karlsruhe, Germany) which was mounted on the lens revolver. A motorized xy-table from Märzhäuser (Wetzlar, Germany) with a motion controller was used to arbitrarily shift the object slide in both x and y direction. For the digital image acquisition a 1.4 Mpixel color digital camera from Soft Imaging System (SIS, Münster, Germany) was used. Our estimates showed that a pixel number larger than 1.4 Mpixel is sufficient for the given magnification. Fig. 1 demonstrates that the optical resolution is sufficient to recognize details in spores like Ulocladium.

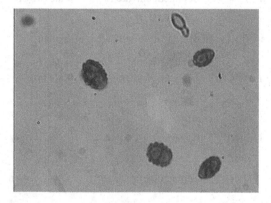

Fig. 1. Image demonstrating the resolution of the optical microscope used. The microscopical image displays spores of Ulocladium. The field of view is 134×100 μm². The sample was prepared by AUA/JenaBios, lens Olympus 60X/0.70. The resolution in this image is 5 μm.

The functions of image acquisition and image storage, movement of the specimen in x and y direction, and auto-focusing in z-direction are controlled by the AnalySIS Pro software from SIS. A pattern of images at any image position can be freely programmed and stored in a macro-code. This holds as well for the number of images to be captured. If necessary it is possible to capture automatically images at different depths of focus around the optimum position. By the automatic shading correction, the effect of an inhomogeneous illumination of the object can be removed.

3.2 The Automatic Probe-Acquisition and Handling System

The following chapter describes the main units and functions of the demonstration set-up realized in the course of the project. A stock of special object slides covered with a sticky layer from Umweltanalytik Holbach [1], (Fig. 2) is kept in a slide storage. A sliding gripper takes the lowest slide in the storage and transports it into the slit impactor from Umweltanalytik Holbach (Fig. 3). The object slides are separated by distance holders with a corresponding recess, in order to avoid sticking between the slides. The distance holder is removed by the same gripper, now moving in opposite direction and depositing the distance holder into a box. The distance holders can be used again when the slide deposit is reloaded.

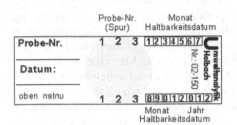

Fig. 2. Object slide of standard size 76×26×1 mm³ with a central sticky layer [1]; Image obtained from Umweltanalytik Holbach

Fig. 3. Slit impactor for collection of airborne particles [1]; Image from Umweltanalytik Holbach

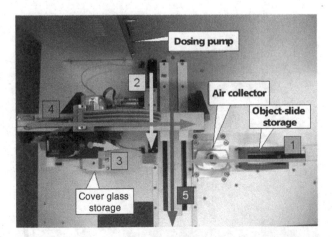

Fig. 4. Top view of the mechanical unit for moving object slides, indicating also the position of the cover-glass storage, the dosing pump for lactophenol, the slit impactor or air collector, and the storage for the object slides. The numbers 1 – 5 indicate the sequences of the movements; axis No. 6 is not shown.

In the slit impactor the air (Fig. 3), potentially containing airborne germs, is guided on the sticky area of the object slide by the air stream generated by a microprocessor controlled air pump. After a few tens of seconds which can be adjusted accordingly, the pump is switched off and the object slide is transported to the pipetting unit driven by the dosing pump (Cavro XL 3000 from Tecan Systems San Jose, Ca, USA. To this aim it has to change its transporting axis and thus its direction of movement. From a thin nozzle one drop of lactophenol is deposited on the sticky area of the object slide which is afterwards transported via the axis crossing to the cover-slip gripper unit. This gripper acts as a low-pressure sucker and takes one cover glass from the deposit and puts it with one edge first on the object slide. Then the cover glass falls down on the object slide and flattens the drop so that it will be distributed all over the sticky area forming a thin layer. In this way the airborne germs collected in the sticky layer are immersed in the lactophenol. In lactophenol living germs get a blue color. The object slide is then transported back to an axis crossing-point where it again changes its direction of movement by 90° and is transported to the xy-table of the microscope which takes over the slide and transports it directly under the lens. The timing of the transportation units, the air and dosing pump is controlled by a distributed multi-axis motion-unit. To this end an additional module was integrated into the AnalySIS Pro software. It controls the manual or automated shift of the xy-table between the image-acquisition position under the lens and the loading position, where the object slide is shifted from the object-slide preparation unit to the xy-table. After the object slide has reached the image acquisition position, the microscope camera then grabs the images at the programmed slide positions after auto-focusing of the microscope lens at each position. The cycle of shifting the xy-table to the defined positions, auto focusing, image acquisition and storage is programmable in a macro-code integrated into the AnalySis Pro software. This can also be done for other procedures like shading correction or image acquisition at different z-positions. After having finished the imaging sequence, the slide is transported away from the xy-table with a special arm

Fig. 5. Prototype set-up showing the dosing pump (arrow 1), several axes, the optical microscope with xy-table (arrow 2), and the digital camera (CC-12, arrow 3). The auto-focusing unit holds the lens (arrow 4).

and falls into a box. When the image grabbing procedure by the microscope unit is still under way, the object-slide preparation unit already starts with the preparation of a new object slide.

The object-slide preparation and manipulation is performed by a hardware controller and by dedicated software written in C++. The transfer from the AnalySIS Pro software to the C++ software and vice versa is controlled by a communication protocol as interface between both software units. Altogether six different mechanical axes have to be handled, not counting the axes of the xy-table (Fig. 4). The unit for object-slide preparation and the expanded microscope are shown in Fig. 5.

4   Image Analysis

Once an image has been taken it is given to the image-analysis unit for further processing. We describe the overall architecture of the system [4][5] and its single components in the next sections.

4.1   The Architecture

The architecture of the system is shown in Figure 6. Objects are recognized in the microscopic image by a case-based object-recognition unit [3]. This unit has a case-base of shapes (case base_1) for fungi spores and determines on a similarity-based inference if there are objects in the image that have a similar shape as the ones stored in the case base. In this case the objects get labeled and are transferred for further processing to the feature-extraction unit. To ensure proper performance of this unit, the general appearance of the shapes of the fungi spores have to be learned. To this end we have developed a semi-automated procedure [3] that allows one to acquire the shape information from the raw image data and to learn groups of shape-cases and general shape-cases. A more detailed description of the case-based object-matching unit can be found in Section 4.2.

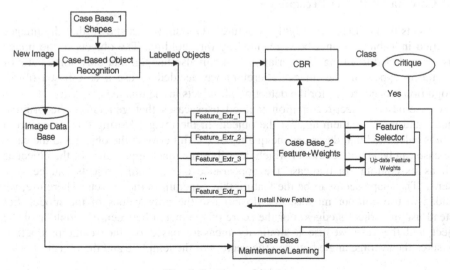

Fig. 6. System architecture

The feature-extraction procedures are based on the knowledge of an expert. Note that a particular application requires special feature descriptors. Therefore not all possible feature-extraction procedures can be implemented into such a system from the beginning. Our aim was to develop a special vocabulary and the associated feature-extraction procedures for application on fungi identification, as described in Section 4.3.

Based on the feature description, the second case-based reasoning unit decides about the type of the fungi spore. This unit employs a prototype-based classifier [11]. It starts its performance on prototypical cases that were selected or created by the expert. It can learn with time the different appearances of the fungi spores. The special features of this unit ensure its proper performance. It can learn the relevant prototypes from the subjectively selected set of prototypes, as well as create new prototypes. It can also learn the importance of the features of the cases. The final result of the system will be the identification of the fungi spores that appear in the image and the number of these spores. This is shown on the display of the system and in a file, together with the date and the time when the data were acquired.

Suppose that fungi species are wrongly identified by the system. Then a case-based maintenance process will start. First it has to be checked by the system developer whether new features have to be acquired for each case, or whether the whole case representation should be updated based on the learning procedures. The feature weights are learnt, as well as a subset of relevant features (see Section 4.4). To acquire new features means that the necessary feature-extraction procedures have to be developed and that for all cases the new features have to be calculated and fed into the existing case description. Therefore we keep the digital images acquired so far in the image-data base. Then the case representation has to be updated as well as the index structure. This ensures that we can come up step-by-step with a system which can describe the variability of the different biological objects that can appear.

4.2 Case-Based Object Recognition

The objects in the image are highly structured. Our study has shown that the images specified in Table 1 cannot be segmented by thresholding. The objects in the image may be occluded touching, or overlapping. It can also happen that only some parts of the objects appear in the image. Therefore we decided to use a case-based object recognition procedure [3] for the detection of objects in the image.

A case-based object-recognition method uses cases that generalize the original objects and matches them against the objects in the image. During this procedure a score is calculated that describes the quality of the fit between the object and the case. The case can be an object model which describes the inner appearance of the object as well as its contour. In our case the appearance of the entire objects can be very diverse. The shape seems to be the feature that generalizes the objects. Therefore, we decided to use contour models. We do not use the gray values of the model, but instead use the object's edges. For the score of the match between the contour of the object and the case we use a similarity measure based on the scalar product. It measures the average angle between the vectors of the template and the object.

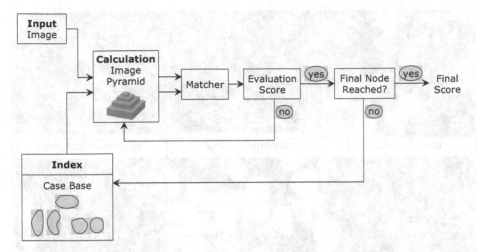

Fig. 7. Principle of case-based object-recognition architecture

4.2.1 Case-Base Generation

The acquisition of the case is done semi-automatically. Prototypical images are shown to an expert. The expert manually traces the contour of the object with the help of the cursor of the computer. Afterwards the number of contour points is reduced for data-reduction purposes by interpolating the marked contour by a first-order polynom. The marked object shapes are then aligned by the Procrustes Algorithm [4]. From the sample points the direction vector is calculated. From a set of shapes general groups of shapes are learnt by conceptual clustering which is a hierarchical incremental clustering method [5]. The prototype of each cluster is calculated by estimating the mean shape [5] of the set of shapes in the cluster and is taken as a case model.

4.2.2 Results for Case-Based Object Recognition

We had a total of 10 images for each class at our disposal. From this set of images two images were taken for the case generation. In these two images there were approx. 60 objects. These objects were labeled and taken for the case generation according to the procedure as described in Section 4.2.1. The result was a data base of cases. These cases were applied to the image for the particular class.

The threshold for the score was set to 0.8. We calculated the recognition rate as the number of objects that was recognized in the image to the total number of objects in the images. Note that the recognition rate can be higher than 100 %, since our procedure also operates in image regions where no objects are present due to background noise. The aim is to set-up the case-based object-recognition unit in such a way that the number of false alarms is low.

The results of the matching process are shown in Figs. 8 and 9. The highest recognition rate can be achieved for the objects Aspergillus niger and Scopularioupsi, since the shape of these objects does vary much. This is also expressed by the number of models, see Table 2. These classes have the lowest number of cases. For those classes where the variation of the shape of the objects is high, the number of the

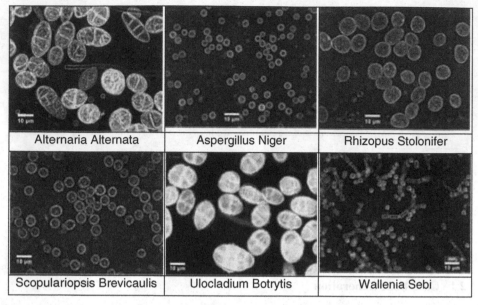

Fig. 8. Recognized objects in the image

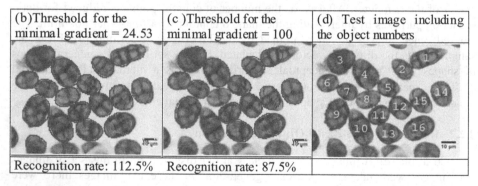

(b)Threshold for the minimal gradient = 24.53	(c)Threshold for the minimal gradient = 100	(d) Test image including the object numbers
Recognition rate: 112.5%	Recognition rate: 87.5%	

Fig. 9. Comparison of the matched objects by applying different thresholds for the minimal gradient

cases is also high. The recognition rate shows that we do not have enough cases to recognize the classes with a good recognition rate (see Ulocladium botrytis and Alternaria alternata). Therefore we need to increase the number of cases. For this task we developed an incremental procedure for the case acquisition in our tool. Objects that have not been recognized well will be displayed automatically for tracing and then the similarity to all other shapes will be calculated. The clustering will be done in an incremental fashion as well [5]. This procedure will ensure that we can learn the natural variation of the shape during the usage of the system.

Table 2. Results of matching

Classes	Number of models	Recognition rate
Alternaria alternata	34	65.9
Aspergillus niger	5	95.2
Rhizopus stolonifer	22	87.7
Scopularioupsi	8	94.5
Ulocladium botrytis	30	77.2
Wallenia sebi	10	90.3

4.3   Case Description and Feature Extraction

We choose an attribute-value pair-representation for the case description. The case consists of the solution which is the type of fungi spores and the features describing the visual properties of the object (see Figure 9). From each recognized object a set of features is extracted. One feature is the case number which represents the shape of the object, the similarity score between the actual shape and the shape in the case base, the size of the object, various gray-level features, and the texture inside the object. For the description of the texture we use our texture descriptor based on random sets described in [6].

4.4   Classification

Our case-based reasoning procedure to recognize spores relies on prototypical-based classification schemes [11]. Usually such schemes are generalized from a set of single cases. Here, we have prototypical cases represented as images that were selected by humans. That means when building our system, we start from the top and have to collect more information about the specific class during the usage of the system. Since a human has selected the prototypical images, his decision on the importance of an image might be biased, and to select only one image might be difficult for a human. He can have stored more than one image as prototypical images. Therefore we need to check the redundancy of the many prototypes for one class before taking them all into the case base. According to this consideration, our system has the following function to fulfill:

- Classification based on the nearest neighbor rule
- Prototype selection by a redundancy-reduction algorithm; Feature weighting to determine the importance of the features for the prototypes
- Feature-subset to select the relevant features from the whole set of the respective domain.

The classification method is based on the nearest-neighbor rule. Since the prototypes are available at the same time, we choose a decremental redundancy-reduction algorithm proposed by Chang [7] that deletes prototypes as long as the classification accuracy does not decrease. The feature-subset selection is based on the wrapper approach [8] and an empirical feature-weighting learning method [9] is used. Furthermore, cross validation is used to estimate the classification accuracy. The prototype selection, the feature selection, and the feature-weighting steps are performed during each run of the cross-validation process. This rule classifies x in the category of its nearest neighbor [10]. More precisely, we call $x'_n \in \{x_1, x_2,...,x_i,...,x_n\}$ a nearest neighbor to x if $\min d(x_i, x) = d(x'_n, x)$, where i = 1, 2, ...n. The nearest neighbor rule chooses to classify x into category C_n where x'_n is the nearest neighbor to x and x'_n belongs to class C_n. For the k-nearest neighbor we require k-samples of the same class to satisfy the decision rule. As a distance measure we use the Euclidean distance. The recognition rate was evaluated on a data base of 50 samples for each class based on cross-validation. The result is shown in Table 3. From that we can conclude that the classification accuracy is higher than the recognition rate for some classes. That means that it is more difficult to recognize the objects that are most likely to be fungi spores than to classify them based on the extracted features.

Table 3. Classification accuracy

Classes	Classification accuracy
Alternaria Alternata	90.4
Aspergillus Niger	95.0
Rhizopus stolonifer	92.0
Scopularioupsi	96.0
Ulocladium botrytis	94.0
Wallenia sebi	92.0

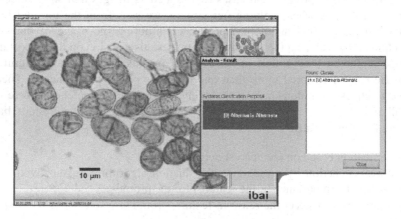

Fig. 10. Screenshot of the final system

A print-out of a result obtained by the system described in this paper is shown in Fig. 10. In the display the operator will find the acquired image in one window and in the other window the determined fungi spores and their total number. The system called Fungi PAD correctly identified the name of the fungi spores and their number.

5 Conclusion

In this paper a system for an automated image acquisition and analysis of hazardous biological material in air is described. It consists of an image-acquisition unit, its sample-handling hardware, and the image-interpretation system. The sample-handling and image-acquisition unit collects the airborne germs, deposits them on an object slide, disperses them with a marker fluid, and takes digital images of the germs in a programmable pattern. The stored images are analyzed in order to identify the germs based on a novel case-based object-recognition method. The case generation is done semi-automatically by manually tracing the contour of the object, by automated shape alignment and by shape clustering, and eventually by prototype calculation. Based on the acquired shape cases, the object-recognition unit identifies objects in the image that are likely to be fungi spores. The further examination of labeled objects is done by calculating more distinct object features, from which a prototype-based classifier determines the kind of fungi spores. After all objects have been classified by their type, the number of one type of fungi spores is calculated and displayed for the operator on the computer screen.

The recognition rate is good enough for on-line monitoring of environments. The final information can be used to determine its contamination with biological hazardous material. It can be used for health monitoring as well as for process control. The described system is the base.

References

1. Mücke, W., Lemmen, Ch.: Schimmelpilze-Vorkommen, Gesundheitsgefahren, Schutzmaßnahmen, Ecomed, Landsberg (2004)
2. Sklarczyk, C., Rieder, H., Rabe, U., Reiß, P., Santraine, B., Lonsdorfer, P.: Entwicklung von Methoden und Verfahren für die bildgebende Erfassung und rechnergestützte Auswertung biologischer Gefahrstoffe, IZFP-Bericht 060618-TW zum Projekt, Optische Bildnahmevorrichtung"; BMWi-Förderkennzeichen 16IN0148
3. http://www.umweltanalytik-holbach.de/index.html
4. Perner, P., Günther, T., Perner, H., Fiss, G., Ernst, R.: Health Monitoring by an Image Interpretation System - A System for Airborne Fungi Identification. In: Perner, P., Brause, R., Holzhütter, H.-G. (eds.) ISMDA 2003. LNCS, vol. 2868, pp. 64–77. Springer, Heidelberg (2003)
5. Perner, P.: Entwicklung von neuartigen Bildanalyse und -auswerteverfahren für luftgetragene biologische Gefahrstoffe, Schlussbericht IBaI, Teilvorhaben, BIOVISION, BMWI- Förderkennzeichen 16IN0147
6. Perner, P., Perner, H., Jänichen, S.: Recognition of Airborne Fungi Spores in Digital Microscopic Images. J. Artificial Intelligence in Medicine AIM 36, 137–157 (2006)
7. Perner, P.: Prototype-Based Classification. Int. J. of Applied Intelligences (in print, 2007)
8. Dryden, I.L., Mardia, K.V.: Statistical Shape Analysis. John Wiley & Sons, Chichester (1998)
9. Jaenichen, S., Perner, P.: Conceptual Clustering and Case Generalization of two-dimensional Forms. Computational Intelligence 22, 178–193 (2006)
10. Perner, P., Perner, H., Müller, B.: Mining Knowledge for Hep-2 Cell Image Classification. Journal Artificial Intelligence in Medicine 26, 161–173 (2002)

11. Chang, C.-L.: Finding Prototypes for Nearest Neighbor Classifiers. IEEE Trans. on Computers C-23, 1179–1184 (1974)
12. Perner, P.: Data Mining on Multimedia Data. LNCS, vol. 2558. Springer, Heidelberg (2002)
13. Wettschereck, D., Aha, D.W.: "Weighting Features", in Case-Based Reasoning Research and Development. In: Aamodt, A., Veloso, M.M. (eds.) Case-Based Reasoning Research and Development. LNCS, vol. 1010, pp. 347–358. Springer, Heidelberg (1995)
14. Aha, D.W., Kibler, D., Albert, M.A.: Instance-based Learning Algorithm. Machine Learning 6, 37–66 (1991)

Geo-Thresholding for Segmentation of Fluorescent Microscopic Cell Images

Tuan D. Pham

Bioinformatics Applications Research Center,
School of Mathematics, Physics, and Information Technology,
James Cook University,
Townsville, QLD 4811, Australia

Abstract. Segmentation is an important research area in image analysis. In particular, effective segmentation methods play an essential role in the computerization of the analysis, classification, and quantification of biological images for high content screening. Image segmentation based on thresholding has many practical and useful applications because it is simple and computationally efficient. Different methods based on different criteria of optimality give different choices of thresholds. This paper introduces a method for optimal thresholding in gray-scale images by mimizing the variograms of object and background pixels. The mathematical formulation of the proposed technique is very easy for computer implementation. The experimental results have shown the superior performance of the new method over some popular models for the segmentation cell images.

Keywords: Segmentation, variograms, bioimaging.

1   Introduction

Thresholding is a simple pattern classification procedure for image segmentation. The key issue of a thresholding algorithm is to choose an optimal threshold value so that the number of misclassified image pixles is kept as low as possible – this is known as the minimum error threshold approach for image segmentation [1,2]. In theory, the optimal threshold value can be determined using Bayes decision rule if the probabilistic distributions of both background and object (foreground) pixels are known [3,4]. However, in most practical cases we do not know the separate distributions but a mixture of both. To handle this problem, we have to make some assumptions about the forms of the distributions and try to determine the optimal threshold according to the principle of statistical decision.

In addition to what we have mentioned above, the inherent difficulty of image segmentation we often encounter is that there are many background pixels that have similar values as those which belong to the object or vice versa. These pixels are usually found in the proximity of the boundaries between the background and the object. These phenomena are particularly common in many bioimaging problems. As a result, these images produce a vague valley in the histograms,

P. Perner and O. Salvetti (Eds.): MDA 2006/2007, LNAI 4826, pp. 15–26, 2007.

which makes it a challenging task for any threshold-based image segmentation methods. There have been numerous attempts for developing image threshold-ing methods for handling different types of images. Many image segmentation methods can be found in several literature reviews over the last three decades [9]. Most general approaches for image segmentation are based on thresholding, clus-ter analysis, edge detection, and region growing. However, a general agreement is that there is no single segmentation method that can be effectively applied to all types of images [10]. Thus, there arises a need for developing new algorithms that may be used for different purposes [11,12,13].

In this paper, a threshold selection method based on a spatial objective cri-terion known as the variogram is introduced for the segmentation of gray-scale images. The mathematical formulation of the variogram is derived from the theory of regionalized variables, which was developed by Matheron [14]. A re-gionized variable is defined as a random variable that is distributed in space. The spatial variability of the regionalized variables can be characterized by both random and structured aspects: (1) they are considered to be erratic in relation to the surrounding variables, and (2) they are spatially related with respect to the distance separating the variables. This spatial structure is called the var-iogram which is a geostatistical function that expresses the spatial relation of the regionalized variables. This conceptual framework is highly applicable to im-age modeling where the pixel values can be thought as being both random and spatially related.

The rest of this paper is organized as follows. In Section 2, we will discuss the concept and procedure for selecting an optimal threshold using the variogram criterion. In Section 3, we will then illustrate and assess the performance of the proposed method by means of the segmented results of fluorescent cell puncta obtained by the proposed and other popular image segmentation methods. The final section is our concluding remarks about the new approach as well as other issues for future development.

2   Threshold Selection from Gray-Level Variograms

Consider a gray-scale image which has intensity values between $[0, T]$ where T is the maximum intensity level. The pixels can be modeled as regionalized variables [14] in the sense that their values are random and they are spatially related. By such hypothesis, the variogram [14,15] of an image is a function which expresses the spatial correlation of the regionalized variables of the image. In probabilistic notation, the image variogram, denoted as $2\gamma(h)$, can be defined as the expected value of the image intensities spatially distributed apart with a distance h:

$$2\gamma(h) = E\{[x_i - x_j]^2\}, \ h_{ij} = h \tag{1}$$

where x_i and x_j are the intensity values of the pixels located at positions i and j of the image respectively, and h is the spatial distance that separates x_i and x_j.

The values of h are taken in any directions in the discrete image. In this study, for the sake of simplicity, h takes the integer values in the horizontal and vertical directions of the image.

The semi-variogram, denoted as $\gamma(h)$, is therefore half of the variogram. The experimental semi-variogram for lag distance h is defined as the average squared difference of values separated by h:

$$\gamma(h) = \frac{1}{2N(h)} \sum_{(i,j)|h_{ij}=h} (x_i - x_j)^2 \qquad (2)$$

where $N(h)$ is the total number of data pairs separated by the distance h.

The behavior of the semi-variogram can be graphically illustrated by the theoretical semi-variogram using the spherical or the Matheron model which is defined as [15]

$$\gamma(h) = \begin{cases} s\left[1.5\frac{h}{r} - 0.5(\frac{h}{r})^3\right] & : \quad h \leq r \\ s & : \quad h > r \end{cases} \qquad (3)$$

where r and s are called the *range* and the *sill* of the semi-variogram respectively.

Figure 1 shows the spherical semi-variogram model defined in (3). When $h = 0$, two samples are taken at the same position and the difference between the two must be zero. When $h > 0$, the two samples move a distance apart and some positive difference between the two values can be expected. As the samples move further apart, the differences should increase accordingly. Ideally when the distance becomes very large and reaches r, the sample values become independent of one another. The semi-variogram $\gamma(h)$ will then become constant at s as the result of the calculation of the difference between the pairs of independent samples.

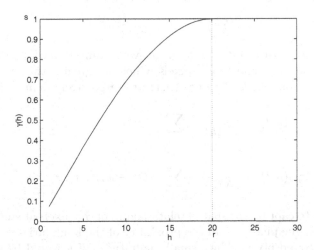

Fig. 1. Example of a semi-variogram – the spherical model with $s = 1$ and $r = 20$

The properties of the semi-variogram can be further explored by again letting h be the distance between two variables x_i and x_j, and by an assumption that the random variables in the random function model has the same mean μ and variance σ^2. These two properties show the relationship between the semi-variogram and the covariance by the following derivation [15]:

$$\gamma(h) = \frac{1}{2}E\{[x_i - x_j]^2\} = \frac{1}{2}E\{x_i^2\} + \frac{1}{2}E\{x_j\}^2 - E\{x_i x_j\}$$
$$= E\{x^2\} - E\{x_i x_j\} = [E\{x^2\} - \mu^2] - [E\{x_i x_j\} - \mu^2]$$
$$= \sigma^2 - C_{ij} \tag{4}$$

where C_{ij} is the covariance of x_i and x_j.

We now consider an image which has *bright* object and *dark* background pixels and wish to segment the image into two classes: object (O) and background (B). Using the concept of the semi-variogram, the total spatial covariance, denoted as $\gamma_I(h)$, of the partition of the image can be expressed as

$$\gamma_I(h) = \gamma_B(h) + \gamma_O(h) \tag{5}$$

where $\gamma_B(h)$ and $\gamma_O(h)$ are the semi-variograms of the background and object pixels respectively.

Our strategy for image segmentation is to select a threshold that separates the image into two groups of homogeneous and spatially related pixels. Thus, a procedure to satisfy this criterion is to minimize the image spatial variance $\gamma_I(h)$. In other words, we seek to minimize the sum of the two spatial variances $\gamma_B(h)$ and $\gamma_O(h)$. Since the segmentation of the two classes is dependent on the variable threshold t and contain pixels with gray values in $[0, t]$ and $[t + 1, T]$ respectively, the spatial variance of I can be rewritten as

$$\gamma_I(h, t) = \gamma_B(h, t) + \gamma_O(h, t) \tag{6}$$

where $\gamma_B(h, t)$ and $\gamma_O(h, t)$ denote the semi-variograms of dark pixels which are less than or equal to t, and bright pixels which are greater than t respectively. The spatial functions $\gamma_B(h, t)$ and $\gamma_O(h, t)$ are respectively defined as

$$\gamma_B(h, t) = \frac{1}{2N(h)} \sum_{(i,j)|h_{ij}=h} (x_i - x_j)^2; \; x_i, x_j \leq t \tag{7}$$

$$\gamma_O(h, t) = \frac{1}{2N(h)} \sum_{(i,j)|h_{ij}=h} (x_i - x_j)^2; \; x_i, x_j > t \tag{8}$$

The purpose is to capture the spatial relationships of the pixels at small distances because most vague pixels are in the proximity of the boundaries of the objects and the background pixels. Thus, small magnitudes of h would be sufficient to model the spatial correlation of the image data. Because each lag distance h may yield a different value for $\gamma_I(h,t)$, we can compute the resultant semi-variogram of the image as the average value of $\gamma_I(h,t)$ by

$$\bar{\gamma}_I(t) = \frac{1}{H} \sum_{h=1}^{H} \gamma_I(h,t) \tag{9}$$

where H is the number of the values of h.

Finally, the spatial optimal threshold value, denoted as t_V, can be determined by searching for the value in the range $[0,T]$ so that $\bar{\gamma}_I(t)$ is minimum. That is,

$$t_V = \arg \min_{0 \leq t \leq T} \bar{\gamma}_I(t). \tag{10}$$

3   Experimental Results

We tested the proposed method with many real fluorescent images of peroxisomes contained in cells. The discrimination and measurement of fluorescent-labeled vesicles using microscopic analysis of fixed cells presents a challenge for biologists interested in quantifying the abundance, size and distribution of such vesicles in normal and abnormal cellular situations. Good image segmentation results will allow the precise quantification of changes to the population of a major organelle, the peroxisome, in cells from normal control patients and from patients with a defect in peroxisome biogenesis.

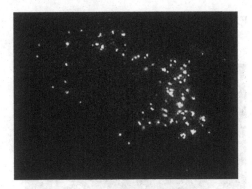

Fig. 2. Original image A

To compare the proposed variogram-based method with some other segmentation methods, we used the Otsu's method [16], the fuzzy c-means (FCM) [17], and the watershed algorithm [2] using the Matlab Image Processing Toolbox to carry out the segmentation of the same images. Two typical images of fluorescent puncta as shown in Figures 2 and 9. These images show the variability of the image database and were selectively collected to test the proposed method.

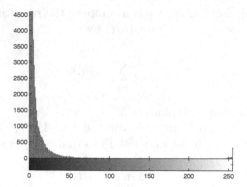

Fig. 3. Histogram of original image A

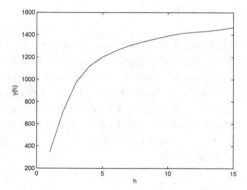

Fig. 4. Semi-variogram of original image A

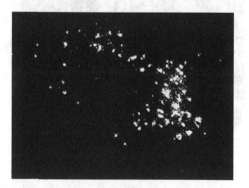

Fig. 5. Otsu-method segmentation of A

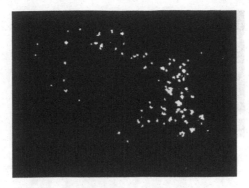

Fig. 6. FCM-based segmentation of A

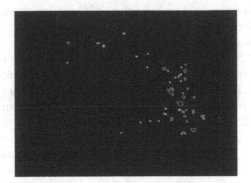

Fig. 7. Watershed-based segmentation of A

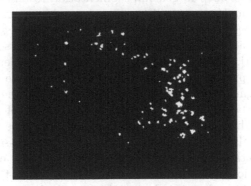

Fig. 8. Variogram-based segmentation of A

The fluoresecence-stained background between and along the boundaries of the spots makes it a difficult task for extracting the correct sizes of the objects [18].

Figures 3, and 10, show the histograms of the original images in Figures 2, and 9 respectively. The two histograms do not show the distinct modes of the

Fig. 9. Original image B

foreground and background pixels, which make it a difficult task for image segmentation. Figures 4, and 11 show the semi-variograms of the original images in Figures 2, and 9 respectively. The behaviors of these image semi-variograms are similar to that of the ideal semi-variogram using the spherical model as defined in (3) and graphically illustrated in Figure 1.

Figures 5, and 12 show the segmentation results of the original images in Figures 2, and 9 respectively by the Otsu methd. Figures 6, and 13 show the segmentation results of the original images in Figures 2, and 9 respectively by the FCM method. Figures 7, and 14 show the segmentation results of the original images in Figures 2, and 9 respectively by the watershed algorithm. Figures 8, and 15 show the segmentation results of the original images in Figures 2, and 9 respectively by the proposed approach. For the proposed variogram-based thresholding, $\bar{\gamma}_I(t)$ was taken as the average of the five semi-variograms having five lag distances ranging from 1 to 5.

It is well-known that the assessment of image segmentation results are not straightforward [7,19,20,21]. In general, segmentation results are considered favorable if the segmented regions are homogeneous and have smooth and spatially accurate boundaries [6]. In addition, the assessment of the segmentation results of biological images are particularly dependent on the experts when classication does not involve [21]. Using this guideline and by visual obervation of biology experts, the results presented in all the figures show that the proposed method provided better segmented images than other methods. The variogram-based thresholding could better recognize the true boundaries of the objects than both Otsu thresholding and FCM-based segmentation. The two latter methods yielded many false positive pixels and resulted in over-segmentation. The watershed-based method provided similar results with those obtained from the proposed segmentation method. However, it can be observed that the segmentation result obtained from the watershed method for the images shown in Figure 2 fails to highlight the larger spots; whereas this visual effect can be noticed in those obtained from the proposed approach. The superior performance of the variogram-based method to the watershed-based method can be better observed in the results for the original images shown in Figure 9 where the watershed algorithm oversegmented the object areas. In general, the variogram-based thresholding gives the segmentation of the spots with more accurate boundaries than

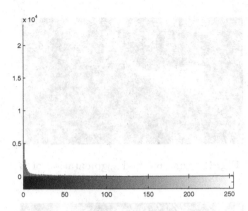

Fig. 10. Histogram of original image B

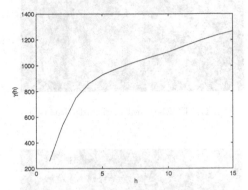

Fig. 11. Semi-variogram of original image B

the other methods and therefore can be effective for the task of cell-imaging quantification.

The main advantages of the proposed method are its abilities to segment images where the sizes of the object and background pixels are heavily unequal, and to assign pixels which are scatteredly distributed over the whole image as the background ones. This type of images are very common in cellular and molecular imaging where the regions of interest occupy only a very small part of the image, and the staining effect is usually scattered over the whole image space. Regarding the computational time, for an image size of 1040×1392, runnning Matlab codes on an average PC, it took about one second for Otsu thresholding; a few seconds for the watershed method, twelve minutes for the FCM-based segmentation; and about seven and two minutes for the variogram-based thresholding taking five ($h=1, \ldots,$ and 5) and three ($h=1, 2,$ and 3) separating distances respectively and both giving similar results for this type of images. It is observed in many cases that using only $h=1$ or $h=1$ and 2 for the geo-thresholding method, good results are obtained for the segmentation of cell images and the running time is dramatically reduced.

Fig. 12. Otsu-method segmentation of B

Fig. 13. FCM-based segmentation of B

Fig. 14. Watershed-based segmentation of B

Fig. 15. Variogram-based segmentation of B

4 Conclusion

A new optimal thresholding method for image segmentation based on the variogram criterion has been discussed. The experimental results have shown its promising application to the segmentation of cell puncta in low-contrast and fluorescent-stained images. Effective segmentation of such biological images helps life-science researchers obtain useful imaging information for downstream analysis including disease diagnosis, treatment, and new drug discovery. The thresholding obtained from the minimization of the variograms is considered optimal in a spatial context, which can be combined with other optimal criteria for generalization or can be used for hysteresis thresholding as a hard-core threshold [1] by taking into account the information at various spatial distances.

Acknowledgments

The author would like to thank Denis I. Crane and Tam Nguyen, both of the Cell Biology Group, the Eskitis Institute, and the School of Molecular and Biomedical Science, Griffith University, for providing the images of fluorescent cell puncta; and Matthew Kerwin, intern of the JCU Bioinformatics Applications Research Centre, for implementing the Matlab codes and carrying out the experiments. This work was supported by JCU 2006 New Staff Grant Awards.

References

1. Petrou, M., Bosdogianni, P.: Image Processing: The Fundamentals. John Wiley & Sons, New York (1999)
2. Gonzalez, R.C., Woods, R.E.: Digital Image Processing, 2nd edn. Prentice Hall, New Jersey (2002)
3. Therrien, C.W.: Decision Estimation and Classification: An Introduction to Pattern Recognition and Related Topics. John Wiley & Sons, New York (1989)
4. Chi, Z., Yan, H., Pham, T.: Fuzzy Algorithms: With Applications to Image Processing and Pattern Recognition. World Scientific, Singapore (1996)
5. Fu, K.S., Mui, J.K.: A survey on image segmentation. Pattern Recognition 13, 3–16 (1981)
6. Haralick, R.M., Shapiro, L.G.: Image segmentation techniques. Computer Vision, Graphics, and Image Processing 29, 100–132 (1985)
7. Pal, N., Pal, S.: A review on image segmentation techniques. Pattern Recognition 26, 1277–1294 (1993)
8. Sankur, B., Sezgin, M.: Survey over image thresholding techniques and quantitative performance evaluation. J. Electronic Imaging 13, 146–165 (2004)
9. Qiao, Y., Hu, Q., Qian, G., Luo, S., Nowinski, W.L.: Thresholding based on variance and intensity contrast. Pattern Recognition 40, 596–608 (2007)
10. Tizhoosh, H.R.: Image thresholding using type II fuzzy sets. Pattern Recognition 38, 2363–2372 (2005)
11. Perner, P.: An architeture for a CBR image segmentation system. J. Engineering Application in Artificial Intelligence 12, 749–759 (1999)

12. Frucci, M., Sanniti di Baja, G.: Object detection in watershed partitioned grey-level images. In: Perner, P., Salvetti, O., Bichindaritz, I. (eds.) ICDM 2006. LNCS (LNAI), vol. 4065, pp. 5–14. Springer, Heidelberg (2006)
13. Frucci, M., Perner, P., Sanniti di Baja, G.: Watershed segmentation via case-based reasoning. In: Weber, R., Richter, M. (eds.) ICCBR 2007. LNCS, vol. 4626, pp. 13-16, 419–432 (2007)
14. Matheron, G.: The theory of regionalized variables and its applications. Paris School of Mines Publication, Paris (1971)
15. Isaaks, E.H., Srivastava, R.M.: An Introduction to Applied Geostatistics. Oxford University Press, New York (1989)
16. Otsu, N.: A thresholding selection method from gray-level histograms. IEEE Trans. Systems, Man, and Cybernetics 9, 62–66 (1979)
17. Bezdek, J.C.: Pattern Recognition with Fuzzy Objective Function Algorithms. Plenum Press, New York (1981)
18. Pham, T.D., Crane, D., Tran, T.H., Nguyen, T.H.: Extraction of fluorescent cell puncta by adaptive fuzzy segmentation. Bioinformatics 20, 2189–2196 (2004)
19. Cinque, L., Foresti, G., Lombardi, L.: A clustering fuzzy approach for image segmentation. Pattern Recognition 37, 1797–1807 (2004)
20. Hoover, A., Jean-Baptiste, G., Jiang, X., Flynn, P.G., Bunke, H., Goldgof, D.B., Bowyer, K., Eggert, D.W., Fitzgibbon, A., Fisher, R.B.: An experimental comparison of range image segmentation algorithms. IEEE Trans. Pattern Analysis and Machine Intelligence 18, 673–689 (1996)
21. Martin, A., Laanaya, H., Arnold-Bos, S.: Evaluation for uncertain image classification and segmentation. Pattern Recognition 39, 1987–1995 (2006)

Image Segmentation by Non-topological Erosion and Topological Expansion

Giuliana Ramella and Gabriella Sanniti di Baja

Institute of Cybernetics "E. Caianiello", CNR, Pozzuoli (Naples), Italy
g.ramella@cib.na.cnr.it, g.sannitidibaja@cib.na.cnr.it

Abstract. A new segmentation method is suggested to distinguish the foreground from the background in gray-level images. The method is based on a 2-step process, respectively employing non-topological pixel removal (non-topological erosion) and topological region growing (topological expansion). The first step is aimed at identifying suitable seeds, corresponding to the objects of interest in the image, while the second step associates to the identified seeds pixels removed during the first step, provided that fusions are not created. Segmentation is accomplished by using also information derived from a lower resolution representation of the image, with the purpose of reducing the number of foreground components to the most significant ones. Some hints regarding extension of the method to color images are also discussed.

1 Introduction

Image analysis deals with the automatic processing of digital images. Whichever is the final goal, a crucial task that has to be faced is image segmentation. This process is necessary to distinguish the foreground from the background and the method to be used depends on problem domain.

A computationally convenient segmentation scheme is based on histogram thresholding. A threshold is determined, based on the histogram of the gray-levels, and all pixels with gray-level larger than the threshold are assigned to one of the two sets (the foreground or the background), while all the remaining pixels are assigned to the other set. In this way, a dichotomy of the image is originated. If the original image is naturally a binary image, e.g., a graphical document including written text, logos and drawings, only two values are used to denote the obtained foreground and background, respectively. Alternatively, when the information carried on by gray-levels are still necessary to further analyze the foreground components, the resulting image will consist of a foreground, whose pixels have the same gray-levels as their homologous pixels in the original image, superimposed on a homogeneous background. A recent survey of about 40 thresholding methods can be found in [1]. Besides categorizing the various methods in terms of the information used to achieve the goal (histogram shape, measurement space clustering, entropy, object attributes, spatial correlation and local gray-level surface), the survey also presents an interesting performance evaluation of several histogram thresholding methods.

P. Perner and O. Salvetti (Eds.): MDA 2006/2007, LNAI 4826, pp. 27–36, 2007.

Histogram thresholding is computationally convenient, but does not take into account spatial information. Thus, it may be difficult to identify a single threshold for image binarization. This happens, for instance, when the same gray-levels characterize pixels that a user would classify, depending on the local context, in some parts of the image as belonging to the background, and in other parts of the image as belonging to the foreground. For these images the threshold should assume different values in different parts of the image, to allow correct assignment of pixels to the foreground and the background, respectively. The best results are obtained by resorting to locally adaptive thresholding methods, which compute a different threshold for each pixel in the image, by means of measures accomplished in suitable pixel neighborhoods (see, e.g., [2-6]).

Besides, histogram thresholding, a number of alternative approaches to image segmentation have been suggested in the literature, such as those based on feature space clustering, edge-detection, or watershed transformation. All these methods have advantages and disadvantages. We do not discuss these alternative methods, since the technique that we suggest in this paper is somehow more related to histogram thresholding than to the other approaches mentioned above. A comprehensive description of these approaches can be found in [7-11].

Our image domain consists of microscope images of cells, where the foreground is perceived as locally darker than the background, consistently through the whole image, even if the same gray-levels can characterize both pixels in areas understood as belonging to the foreground and pixels in areas perceived as belonging to the background. Gray-levels are in the range [0-255] and we assume that 0 corresponds to the lightest gray-value (white) and 255 to the darkest one (black).

Similarly to histogram thresholding methods, we use the histogram of the gray-levels to identify the range of gray-levels that we regard as certainly characterizing the foreground and the background for the specific application. Differently from histogram thresholding methods, the not yet assigned pixels, i.e., those whose gray-levels are in between the gray-levels regarded as definitely characterizing foreground pixels and background pixels, undergo an iterated 2-step process, respectively employing non-topological pixel removal (called in the following *non-topological erosion*) and topological region growing (called *topological expansion*). The number of iterations is fixed a priori as the number n into which the range of gray-levels of the not yet assigned pixels is sampled. At the k-th iteration, if g_k is the maximum gray-level value that we consider, only the not yet assigned pixels with gray-level up to g_k are examined. Non-topological erosion is done to orderly remove pixels with gray-level up to g_k, provided that they belong to the border of foreground components, in order to split these components into a number of disjoint sets (the seeds). Topological expansion is then accomplished by adding again to the seeds those pixels, among the previously removed ones, which do not create fusions of regions. If at the k-th iteration all the removed pixels can be added again, this means that no separation of foreground components occurs in correspondence of gray-levels until g_k. In any case, the process continues and the possible foreground separations at a higher gray-level are searched. Thus, our method can be classified as a multi-threshold method, since foreground components can be separated at different gray-level values. A peculiarity of our method is that we avoid an excessive reduction of the size of foreground components, due to the accomplished topological expansion that restores all removed pixels, except those in the fusion areas. Another feature of our method is that we use information derived from a lower resolution representation of the image to reduce the number of foreground components to the most significant ones.

2 The Method

In the following, we will describe our method by using as running example the image G shown in Fig. 1 with its histogram. For visualization purposes, the running example has a small resolution (128×128), but we actually process larger size images.

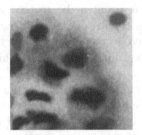 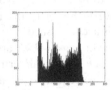

Fig. 1. Running example and gray-level histogram

Let g denote any pixel of G as well as its associated gray-level. We need to fix two thresholds, θ_i and θ_f, depending on the gray-level distribution in G. The value of θ_i should be such that all pixels with gray-level $g \leq \theta_i$ can definitely be interpreted as background pixels in the specific application. The value θ_f should be such that all pixels with gray-levels $g > \theta_f$ certainly belong to the foreground. In general, θ_i and θ_f can be set in correspondence with the leftmost and the rightmost valleys in the histogram, by assuming that the first and the last peaks of the histogram, including the lightest and the darkest gray-levels respectively, definitely allow to identify portions of the background and of the foreground.

If the histogram is mainly bimodal, a generally large valley separates the two main peaks and the two thresholds can be set at the two extremes of the valley. For a multimodal histogram, the thresholds can be shifted towards more internal valleys, depending on user's needs. In fact, a too small value of θ_i could cause the detection of a number of noisy components, erroneously interpreted as belonging to the foreground. In turn, the value of θ_f should be set in such a way to prevent excessive foreground fragmentation, which would split foreground components perceived as individual entities into a number of meaningless components. In the current version of the method, the selection of θ_i and θ_f is not automatic, but is based on a priori knowledge on the characteristics of the input image. Pixels with gray-level value g, $\theta_i < g < \theta_f$, can be foreground pixels or background pixels in different parts of the image. For the running example, the two thresholds have been set to $\theta_i = 176$ and $\theta_f = 206$.

2.1 Non-topological Erosion

The range of gray-levels between θ_i and θ_f is sampled into n groups. Each group gathers δ successive gray-levels. In our case, it is $\delta=5$. This choice is motivated by a compromise in computational efficiency (the larger is δ, the smaller is n and, hence, the number of iterations of the process) and quality of the result (with a small δ, also perceived foreground components which are remarkably lighter with respect to other foreground components can be singled out. In turn, if δ is large, lighter foreground components risk to be lost, since no seeds would be detected in correspondence with them at any iteration to allow their recovery after non-topological erosion). For the running example it results $n=6$ and, hence, six iterations of the process are accomplished. At the first iteration, pixels with gray-level g, $\theta_i<g<\theta_i+\delta$, are examined. In general, at the k-th iteration, pixels with gray-level g, $\theta_i<g<\theta_i+k\times\delta$, are examined.

To separate foreground components that result to be merged in a fusion area characterized by a gray-level value, which is smaller than $\theta_i+k\times\delta$, the pixels placed in the fusion area have to be identified and have to be assigned to the background (i.e., their gray-levels have to be set to the background value θ_i). To this purpose, non-topological erosion is accomplished.

Pixels with gray-level g, $g<\theta_i+k\times\delta$, are examined in increasing gray-level order and those having at least one horizontal/vertical neighbor in the background, i.e., at least a neighbor with gray-level equal to θ_i, are temporarily assigned to the background, regardless of the topology changes possibly caused by this operation. Non-topological erosion is carried on until border pixels are found. Though border pixels are removed independently of whether connectedness is preserved in their neighborhoods, the number of connected components of pixels with gray-level greater than $\theta_i+k\times\delta$ present in the neighborhood of any removed pixel is counted by using the connectivity number C8 introduced in [12]. If for a pixel p it results C8>1, this indicates that p was actually a fusion point among foreground components. Thus, a fusion flag f_c, initially set to zero, can be suitably changed to record the event.

At the end of non-topological erosion, the flag f_c is examined. If $f_c=0$, no fusion occurred at the current iteration, and all removed pixels can be recovered to preserve the information contents of the image. In turn, if $f_c\neq0$, some fusion actually occurred. In this case, only the removed pixels that are far from the fusion area can be recovered. In any case, before proceeding to the successive iteration of non-topological erosion, topological expansion is performed, to restore size and shape of foreground components, which have been modified by non-topological erosion.

A crucial difference with respect to a thresholding method that would assign to the background all pixels with gray-level g, $g<\theta_i+k\times\delta$, is that our criterion removes only border pixels. Thus, pixels characterized by gray-level g, $g<\theta_i+k\times\delta$, possibly present within a darker component are not assigned to the background, so that spurious background components are not created. Another advantage of removing pixels by our method is that we can record the list L of removed pixels in the order they have been removed. This will play a useful role during topological expansion.

In Fig. 2, the images resulting after the 5-th iteration and the 6-th iteration of non-topological erosion are shown, where black and gray respectively denote the seeds and the pixels (temporarily) assigned to the background. We note that the individual

Fig. 2. Results after the 5-th iteration, left, and the 6-th iteration, right, of non-topological erosion

seed in the bottom-left part of the image in Fig. 2 left, is split in two seeds in the image in Fig. 2 right.

2.2 Topological Expansion

During the k-th iteration of topological expansion, recovery of pixels removed at the same iteration of non-topological erosion is done in such a way to maintain separated the components whose seeds have been singled out. The process is straightforward if no fusion occurred at the k-th iteration, since all pixels stored in L are just newly assigned their initial gray-level in the image G.

A more tricky process is necessary if $f_c \neq 0$, i.e., if some fusion occurred. In this case, the pixels recorded in L are examined and set to their initial value in G in the opposite order with respect to the order in which they were inserted in the list. In fact, the pixels that were removed last are the first ones that can be recovered, being the closest ones to the seeds identified by non-topological erosion. The number, C8, of components of foreground pixels (i.e., pixels with gray-level greater than θ_i) is counted in the neighborhood of each pixel p restored from L in the image G. If C8>1, p is assigned on G its initial gray-level, but it is also marked in G as a fusion point.

The set of pixels marked as fusion points is generally not sufficient to separate foreground components associated to distinct seeds. Thus, a marker propagation process is performed to identify all pixels in the fusion areas. To this aim, after all pixels of L have been processed and restored in G, all the pixels recovered from L and neighbors of marked pixels are also marked as fusion points. Obviously, by indefinitely iterating the marker propagation process as far as recovered pixels can be marked, the detected set of fusion points will generally include a larger number of pixels with respect to those whose removal is really necessary to maintain separated the foreground components associated to distinct seeds. In some cases, all the removed and recovered pixels surrounding a seed result to be marked as fusion points at the end of marker propagation. Therefore, to assign to the background a limited number of pixels, a suitable process is performed to remove the marker from pixels far from the actual fusion areas. Pixels still marked as fusion points are finally set to θ_i, so obtaining the separation of foreground components at the k-th iteration.

The images resulting after the 5-th iteration and the 6-th iteration of topological expansion are shown in Fig. 3, where black and gray denote foreground and background, respectively.

 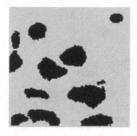

Fig. 3. Results after the 5-th iteration, left, and the 6-th iteration, right, of topological expansion

The iterated application of non-topological erosion and topological expansion allows us to separate foreground components whose average gray-levels differ, or have similar average gray-levels but are separated by portions of the background characterized by different average gray-levels. This is not possible if a single image thresholding is accomplished. We point out that once a foreground component, lighter than other components, has been singled out at a given iteration, this component will be preserved even when, at some successive iteration, non-topological erosion, will completely remove such a component for a suitable value of $\theta_i + k \times \delta$. In fact, the first pixel recovered by topological expansion in correspondence of such a component will be characterized by C8=0, being an isolated pixel, and all successively recovered pixels will be characterized by C8=1, so that none of the recovered pixels of the component will be marked as fusion point. This is the case, for example, for the foreground components in the top right and top left portions of the images in Fig. 3, for which no seeds were detected (see Fig. 2). Actually, those components were singled out at the first iteration. During all successive iterations, they are completely removed by non-topological erosion and recovered by topological expansion.

2.3 Using Lower Resolution Image Representation

Though the foreground is perceived as consisting of sets of rather homogeneously dark pixels, gray-level variations not necessarily meaningful but still enough to split a foreground component in more than one seed are possible. This is particularly true when the value selected for δ is rather small, as it is in our case. The negative effect of a small δ is that the number of seeds results larger than the number of perceived foreground components and some components obtained at the end of the topological expansion are scarcely meaningful. Since increasing δ could prevent us to single out some desired lighter foreground components in correspondence of which no seeds would be found, we suggest an alternative criterion to remove the seeds that do not correspond to meaningful regions, based on the use of a lower resolution representation of the image.

It is well known that if representations at different resolution of a gray-level image are available, only the most significant regions will be perceived at all resolutions. Regions with lower significance, which can be interpreted as fine details, are perceived only at sufficiently high resolution. Thus, if non-topological erosion and topological expansion are accomplished on the image at full resolution as well as on a lower resolution representation of the image, the components detected at full resolution, but that are not present at lower resolution, can be regarded as less significant with respect to components found in both images.

To build a lower representation of the image, we use a decimation process that associated to each 2×2 set of pixels in the full resolution image a single pixel p' in the lower resolution image G'. The gray-level of p' is computed in terms of the gray-level of one of the pixels, say p, placed in block of 2×2 pixels in the full resolution image, as well as of the gray-levels of the eight neighbors of p. Since there are four possible ways to divide G into blocks of 2×2 pixels, the pixel p can actually be in any of the four positions of a block. For the same reason, the eight neighbors of p can be present in one or more of the four possible blocks. Horizontal/vertical neighbors of p belong to two different blocks, while each diagonal neighbor of p belongs to one block only. Thus, we can weight the gray-levels of the neighbors of p with values 2, for horizontal/vertical neighbors, and 1, for diagonal neighbors. Since p is present in all four blocks, the gray-level of p can be weighted as 4. In this way, we can compute the gray-level of p' in a way almost independent of the division of G into blocks of 2×2 pixels. Obviously, gray-levels have to be rescaled to the range [0-255]. To compute G', pixels in even rows and columns of G are considered as the bottom right pixels in the 2×2 blocks corresponding to single pixels in G'. It is immediate to see that, given a pixel p' with coordinates (i,j) in G', the coordinates of the four pixels in the corresponding 2×2 block in G can be easily determined.

If a representation of G at even lower resolution is desired, the same process can be repeated on G'. In practice, a pyramid representation of G can be built. See, for example, [13,14] where multi-scale pyramid representation was discussed. There, a preliminary version of the segmentation method here illustrated was used to segment the full resolution image so as to build a topology-preserving pyramid.

Selection of the lower resolution representation to be used in order to reduce the number of components detected at full resolution depends on problem domain. If the objects perceived in the input image by the expert have large size, a quite lower resolution will still reasonably reflect the properties of G and the most important components will still be those detected at the low resolution representation. Of course, if G includes small size foreground components, resorting to a very low resolution image is not advisable, since the input image would be represented in a too coarse way. For the running example, we consider the image G' with resolution 64×64 pixels as adequate to the aim. See Fig. 4 top, showing the full resolution image and the 64×64 lower resolution image.

Non-topological erosion and topological expansion are accomplished on both G and G'. Components identified in G, such that the relative 2×2 blocks all correspond to background pixels in G' are removed (i.e., their pixels are assigned to the background).

In Fig. 4 bottom, the components found in G and G' at the end of the process are shown on a gray background. A lighter gray tone is used to mark in the full resolution image the components that will be assigned to the background, due to the information derived from the lower resolution image.

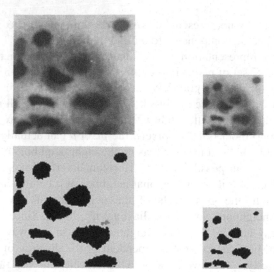

Fig. 4. Top: the 128×128 image, left, and its 64×64 lower resolution representation, right. Bottom: the components found at the end of the process for the 128×128 image, left, and the 64×64 image, right.

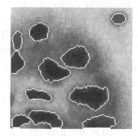

Fig. 5. The segmentation obtained for the running example

In Fig. 5, the segmentation of the running example is shown. The white lines, superimposed on the original image, identify the borders of the components of the foreground that have been singled out.

3   Extending the Method to Color Images

In this section, we briefly discuss a possible way to extend our method to color images. Starting from an input color image C, the three gray-level images in the (RGB) color space are computed. The segmentation process is applied to each of these images, by selecting for each of them the proper values for θ_i and θ_f.

A simple way to combine the three resulting images, is to binarize them and compute the OR image to obtain the foreground components of the input image. The

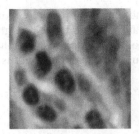 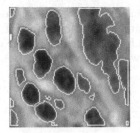

Fig. 6. A color input image, left, and its segmentation, right

effect is shown in Fig. 6, where the white lines superimposed on the input image identify the borders of the detected foreground components.

Alternatively, the three binary images could be added to each other, producing an image with values ranging from 0 to 3. Pixels with value 3 belong to components identified in all the segmented images, pixels with value 2 or 1 belong to components detected in two or just one segmented image. Obviously, pixels with value 0 have been classified as background pixels in the three segmented images. Pixels belonging to each connected component of 3's, 2s and 1's could be labeled with the average color of the corresponding region in the input image. We have noted that, if this method is adopted, some region merging becomes necessary. We are currently working on this problem.

4   Conclusion

In this paper we have presented a segmentation scheme for gray-level images based on two processes: non-topological erosion and topological expansion. The method is semi-automatic since the user is requested to select, based on the gray-levels histogram, two thresholds, θ_i and θ_f, to assign to the background and to the foreground pixels definitely perceived as belonging to these two sets. The remaining pixels undergo iterated non-topological erosion and topological expansion. To this aim, the range of gray-levels of the not assigned pixels is sampled into n intervals including a small number δ of successive gray-levels. Non-topological erosion will remove, at the k-th iteration, only pixels, which are adjacent to background pixels and have gray-level smaller than $\theta_i + k \times \delta$. These pixels will be recovered by topological expansion, provided that they are not found in fusion areas.

Advantages of the method with respect to a thresholding method that would assign to the background all pixels with gray-level smaller than a given threshold, is that during non-topological erosion pixels characterized by small gray-levels, possibly present within a darker component, are not assigned to the background and spurious background components are not created. Moreover, foreground components whose average gray-levels differ, or foreground components with similar average gray-levels but separated by portions of the background characterized by different average gray-levels, can be singled out.

To reduce the number of detected components to the most significant ones, we have also used information derived from a lower resolution representation of the input image. In particular, the lower resolution image undergoes non-topological erosion and topological expansion, and components found in the full resolution image that are not also identified in the lower resolution image are assigned to the background.

We have also suggested a possible extension of the method to color images. In this respect, our work is at a preliminary stage and further research activity is planned for the future.

References

1. Sezgin, M., Sankur, B.: Survey over image thresholding techniques and quantitative performance evaluation. Journal of Electronic Imaging 13(1), 146–165 (2004)
2. Kamel, M., Zhao, A.: Extraction of binary character/graphics images from grayscale document images. Graphical Models Image Processing 553, 203–217 (1993)
3. Nakagawa, Y., Rosenfeld, A.: Some experiments on variable thresholding. Pattern Recognition 113, 191–204 (1979)
4. Deravi, F., Pal, S.K.: Gray level thresholding using second-order statistics. Pattern Recognition Letters 1, 417–422 (1983)
5. Yan, F., Zhang, H., Kube, C.R.: A multistage thresholding method. Pattern Recognition Letters 26, 1183–1191 (2004)
6. Huang, Q., Gao, W., Cai, W.: Thresholding technique with adaptive window selection for uneven lighting image. Pattern Recognition Letters 26, 801–808 (2005)
7. Pal, N.R., Pal, S.K.: A review on image segmentation techniques. Pattern Recognition 26(9), 1277–1294 (1993)
8. Pham, D.L., Xu, C., Prince, J.L.: Current methods in medical image segmentation. Annual Review of Biomedical Engineering 2, 315–337 (2000)
9. Lucchese, L., Mitra, S.K.: Color Image Segmentation: A State-of-the-Art Survey, "Image Processing, Vision, and Pattern Recognition". Proc. of the Indian National Science Academy (INSA-A), New Delhi, India 67 A(2), 207–221 (2001)
10. Freixenet, J., Muñoz, X., Raba, D., Martí, J., Cufí, X.: Yet Another Survey on Image Segmentation: Region and Boundary Information Integration. In: Heyden, A., Sparr, G., Nielsen, M., Johansen, P. (eds.) ECCV 2002. LNCS, vol. 2352, pp. 408–422. Springer, Heidelberg (2002)
11. Soille, P.: Morphological Image Analysis – Principles and Applications, 2nd edn. Springer, Heidelberg (2003)
12. Yokoi, S., Toriwaki, J.I., Fukumura, T.: An analysis of topological properties of digitized binary pictures using local features. Comput. Graphics Image Process. 4, 63–73 (1975)
13. Ramella, G., Sanniti di Baja, G.: Detecting foreground components in grey level images for shift invariant and topology preserving pyramids. In: Campilho, A., Kamel, M. (eds.) ICIAR 2004. LNCS, vol. 3211, pp. 57–64. Springer, Heidelberg (2004)
14. Ramella, G., Sanniti di Baja, G.: Grey level image components for multi-scale representation. In: Sanfeliu, A., Martínez Trinidad, J.F., Carrasco Ochoa, J.A. (eds.) CIARP 2004. LNCS, vol. 3287, pp. 574–581. Springer, Heidelberg (2004)

Variability Analysis of the Large-Scale Structure of Interphase Chromatin Fiber Based on Statistical Shape Theory

Siwei Yang[1], Sandra Götze[2], Julio Mateos-Langerak[2], Roel van Driel[2], Roland Eils[1], and Karl Rohr[1]

[1] Biomedical Computer Vision Group, Dept. Theoretical Bioinformatics, DKFZ Heidelberg, and University of Heidelberg, IPMB, Dept. Bioinformatics and Functional Genomics, Im Neuenheimer Feld 267, D-69120 Heidelberg, Germany
[2] Swammerdam Institute for Life Sciences (SILS), University of Amsterdam, BioCentrum Amsterdam, Kruislaan 318, 1098 SM Amsterdam, The Netherlands
s.yang@dkfz.de

Abstract. The relationship between geometric folding of the chromatin fiber and genome function is a key issue in cell biology. We propose different approaches based on statistical shape theory to investigate the geometric variability of chromatin folding in nuclei of interphase human fibroblasts. Our main purpose is to assess the degree of variability of folding of the chromatin fiber, measured by fluorescent in situ hybridization, using BAC probes in combination with 3D confocal microscopy. We employ point-based registration, the complex Bingham distribution, generalized Procrustes method, and the Kendall spherical coordinate system. The approaches have been applied using 337 3D multi-channel microscopy images. We have analyzed the geometric structure formed by gene-rich highly expressed genomic regions and areas that are gene-poor and have a low transcriptional activity. It turned out that the structure formed by these genomic regions exhibit high shape variation, however, most of them can be characterized by a non-uniform shape distribution.

1 Introduction

The common model of the 3D structure of chromatin assumes that the DNA folds around histone octamers, forming arrays of nucleosomes in a 10 nm fiber, which folds into 30 nm diameter chromatin filament. Remarkably, little is known about higher order folding, despite the fact that the 3D organization of the chromatin fiber plays an important role in the control of gene expression [1]. In this work we are interested in the 3D geometric properties of large-scale chromatin fiber of interphase cells. The general motivation consists in relating geometric information to genome function, in order to obtain a better understanding of how the large-scale chromatin structure affects gene regulation in normal and abnormal cells (for recent surveys on this issue we refer to [2] and [3]). The main

P. Perner and O. Salvetti (Eds.): MDA 2006/2007, LNAI 4826, pp. 37–46, 2007.

purpose of our work is to analyze the variability of the 3D geometric structure formed by different genomic regions identified by fluorescent in situ hybridization (FISH) with bacterial artificial chromosome (BAC) probes. We have acquired 337 three-color 3D confocal microscopy images of nuclei of human fibroblasts in which three genomic regions were FISH labelled on the q-arm of chromosome 1. In this way each image contains a cell nucleus with three spots, representing three genomic regions on the same chromosome, that form a triangle, denoted as BAC-triangle (see Fig. 1, top left and bottom). We analyze the gene-rich and highly expressed genomic regions (called ridges [4]) and the gene-sparse genomic regions showing low gene expression (called anti-ridges). To assess the variability of the structures labelled by the BACs we propose different approaches.

Prior to a statistical evaluation we first apply 3D point-based registration to transform the BACs onto the x-y plane. The purpose is to normalize the data and to reduce the dimensionality of the problem from 3D to 2D. Second, we perform statistical shape analysis to evaluate the shape variability of the datasets. Our analysis is based on the following two approaches: The complex Bingham distribution model [5], which involves one parameter that characterizes the degree of variability of the data, and the generalized Procrustes method [6], which captures the dominant variation of the data. In addition, we employ Kendall's spherical coordinate system [6] to visualize the shape distribution of the BAC-triangles. According to our knowledge statistical shape analysis has not yet been used to assess the variability of the 3D structure of chromatin fibers.

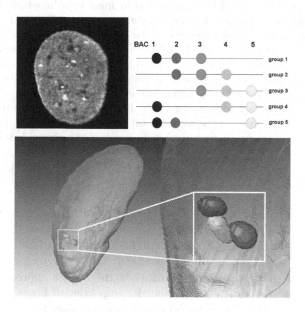

Fig. 1. Top left: One section of a 3D original microscopy image of a cell nucleus; Top right: Relation between BACs and groups of datasets; Bottom: 3D visualization of a cell nucleus and three BACs

2   3D Image Data

Our study is based on 337 3D microscopy images of human fibroblasts in which three specific positions on the chromatin fiber of individual chromosomes have been labeled by fluorescence in situ hybridization (FISH). The multichannel images have a resolution of 512 x 512 x 100 voxels. The data was acquired in five groups according to the scheme sketched in Fig. 1, top right. In subsequent groups there is an overlap of two BACs. For example, within group 1 the positions 1,2,3 are labeled, whereas in group 2 these are positions 2,3,4. In total, we have acquired 337 datasets that are divided in 10 groups: 5 groups for a ridge and 5 groups in an anti-ridge. Each group depicts a specific BAC-triangle measured in about 30 cells. Generally, two BAC-triangles can be observed in a nucleus, since the cells are diploid. However, in ca. 60% of our real image data only one BAC-triangle can be identified unambiguously. In total, we thus have 457 BAC-triangles for our analysis. After applying threshold-based segmentation we have computed the center of mass for each BAC labeled site with sub-voxel resolution and positions have been corrected for chromatic aberration.

Besides the real datasets we have also generated two sets of simulated data which serve as reference datasets. First, we created 50 triangles with low variability. The vertices of the triangles are isotropic normally distributed $N(\boldsymbol{\mu}, \sigma)$, where $\sigma = 0.05$ and the mean side length of the triangles is 0.82. We denote this dataset by "stable triangles". Also, we created a dataset "random triangles" which consists of 50 triangles, whose vertices are uniformly distributed within a unit cube.

3   Methods

3.1   Point-Based Rigid Registration

Prior to applying different techniques from statistical shape theory, we employ 3D point-based rigid registration (translation, rotation) to transform all 3D BAC-triangles onto the x-y plane (reference system). 3D point-based registration can be formulated as follows. Given k source points $\mathbf{p_i}$, and k target points $\mathbf{q_i}$, the task is to find a rigid transformation $\mathbf{R}$ such that $\sum_{i=1}^{k} \|\mathbf{q_i} - \mathbf{p_i} \circ \mathbf{R}\|^2$ is minimized. To register BAC-triangles ($k = 3$) onto the x-y plane, we arbitrarily selected a triangle in this plane as the target structure, and applied the algorithm of Horn [7]. After registration, each vertex of the triangles can be represented by a 2D coordinate or a complex number.

Note, that originally the BAC-triangles are labeled clockwise or counterclockwise. However, after point-based registration, there is exclusively one kind of labeling order, i.e. either only clockwise or only counter-clockwise, because the counter-clockwise order and the clockwise order can be transformed to each other by a 3D rotation. This is called removing the reflection shape [6].

3.2   Complex Bingham Distribution

After having transformed the BAC-triangles onto the x-y plane, we use the complex Bingham distribution to model the shape distribution. This technique provides an elegant framework for the analysis of 2D shape data [6]. The main advantage is that only one parameter is involved and that this parameter characterizes the degree of shape variability, e.g., it indicates whether the shape distribution of triangles has the tendency to be uniform. Below, we introduce the complex Bingham distribution in the context of our application.

Given a set of n triangles (number of vertices $k = 3$), which have been transformed onto the x-y plane. Each triangle can be represented by a 3D complex vector $\tilde{\mathbf{z}}_i = (\tilde{z}_{i1}, \tilde{z}_{i2}, \tilde{z}_{i3})$, where $\tilde{z}_{ij} \in \mathbb{C}, i = 1, ..., n$, and $j = 1, 2, 3$. A central issue is to examine whether the shape distribution of these triangles is uniform. In this case the shape of the triangles is random. First we have to remove undesirable effects from scaling and translation. To perform this step we need a special transformation (for details we refer to [6]). The transformed triangles are represented by 2D complex vectors $\mathbf{z}_i = (z_{i1}, z_{i2})$. The complex Bingham distribution has the following rotation-invariant probability density function:

$$f(\mathbf{z}) = c(\mathbf{A})^{-1} \exp(\mathbf{z}^* \mathbf{A} \mathbf{z}) \tag{1}$$

where $\mathbf{A}$ is a $(k-1) \times (k-1)$ Hermitian matrix, $c(\mathbf{A})$ is the normalizing constant, and $\mathbf{z}^*$ represents the complex conjugate of the transpose of $\mathbf{z}$. In the case of triangles ($k = 3$) $\mathbf{A}$ has two distinct eigenvalues λ_1 and λ_2 where $\lambda_1 > \lambda_2$. In order to investigate the form of the shape distribution, we need to determine λ_1 and λ_2, and examine whether λ_1 and λ_2 are approximately zero. If this is the case, then the triangles tend to have a uniform distribution in shape space. However, in our application the eigenvalues generally cannot be close to zero. The reason is, that for a uniform shape distribution both labeling orders for triangles must be included, however, 3D triangles only have one labeling order.

The eigenvalues of $\mathbf{A}$ can be computed based on:

$$c(\mathbf{A}) = 2\pi^2 \sum_{j=1}^{k-1} a_j \exp(\lambda_j), \qquad a_j^{-1} = \prod_{i \neq j}^{k-1} (\lambda_j - \lambda_i) \tag{2}$$

Note, that the Bingham distribution remains unchanged if a constant is added to all eigenvalues. The consequence is that for the λ_i there is no unique solution. Fortunately, this non-uniqueness can be conveniently removed by setting the largest eigenvalue to zero without lost of generality. In our case, we set $\lambda_1 = 0$, which implies that the second eigenvalue λ_2 is negative. Then only one parameter remains, which makes our analysis easier. Hence

$$c(\mathbf{A}) = 2\pi^2 \left(\frac{1}{-\lambda_2} + \frac{\exp(\lambda_2)}{\lambda_2} \right) \tag{3}$$

λ_2 is usually estimated by means of maximum-likelihood estimation (MLE). First let $\mathbf{S} = \sum_{i=1}^{n} \mathbf{z}_i \mathbf{z}_i^*$ be the $(k-1) \times (k-1)$ complex sum of squares and products matrix. In the case of triangles ($k = 3$) **S** has two positive and distinct eigenvalues, i.e. $l_1 > l_2 > 0$. Note that $l_1 + l_2 = n$. The log-likelihood for the data reads: $L = l_2\lambda_2 - n\log[c(\mathbf{A})]$ where l_2 is the smaller eigenvalue of the matrix **S** defined above.

Test of Uniformity. To answer the question whether the data has a random shape, i.e., whether the BAC-triangles have uniform shape distribution, we perform the following statistical test. Generally, the shape space of 2D triangles is a spherical space instead of an Euclidean space. Its southern hemisphere contains all triangles with clockwise labeling, whereas all triangles with counter-clockwise labeling are located on the northern hemisphere. However, the shape space of 3D triangles consist of just one hemisphere [6], since 3D triangles have only one kind of labeling as mentioned above. Thus standard methods of directional statistics, which are particularly designed for statistics of spherical data, are not suited. However, the uniformity on the full sphere implies the uniformity on its both hemispheres. Therefore, one possible solution to overcome this drawback is to map half of the 3D triangles onto the other hemisphere. If the mapped dataset is uniform, then the original one is also uniform. To perform the mapping, for each triangle we randomly assign either its original or its reflected shape as input data. Using this scheme half of the data are located on the northern hemisphere and half of the data are located on the southern hemisphere. Subsequently we apply a statistical test on the spherical data as described by Mardia and Jupp [8]. First we need to establish the sum of squares and products matrix $\hat{\mathbf{S}}$, where the corresponding eigenvalues are $\hat{l}_1$ and $\hat{l}_2$. The test statistic $F = 3(\hat{l}_1 - \hat{l}_2)^2/n$ has a chi-squared distribution, i.e. $F \sim \chi_3^2$. This value can be used to determine whether the data is uniform, which is the case for large values of F at a certain significance level, e.g., for the upper 1% quantile of χ_3^2 we have the value 11.34.

3.3 Generalized Procrustes Method

Apart from the evaluation based on the complex Bingham distribution we also investigate the dominant shape variation of BAC-triangles. The generalized Procrustes method uses principal components to characterize the main tendency of structural variability (the term "generalized" indicates, that there are more than two objects involved).

First, it is necessary to compute the full Procrustes estimate of the mean shape [9] for a set of triangles. Afterwards, one can examine how the triangles vary with respect to the mean shape. For this purpose we take advantage of principal components analysis (PCA) of the Procrustes residuals [6]. Let the real vectors $\mathbf{r}_i, i = 1, ..., n$ be the Procrustes residuals, and **M** be the sample covariance matrix of $\mathbf{r}_i$, i.e. $\mathbf{M} = \frac{1}{n}\sum_{i=1}^{n}(\mathbf{r}_i - \bar{\mathbf{r}})(\mathbf{r}_i - \bar{\mathbf{r}})^T$ where $\bar{\mathbf{r}} = \frac{1}{n}\sum \mathbf{r}_i$. The orthonormal eigenvectors of **M** denoted by γ_i, are the principal components (PCs) of **M** with corresponding eigenvalues λ_i. The percentage of variability captured by the ith PC is $100\lambda_i^2/\sum\lambda_i^2$. The effect of the i-th PC can be visualized by adding **r** on the mean shape, where $\mathbf{r} = \bar{\mathbf{r}} + c\lambda_i^{1/2}\gamma_i$ for a range of values of the standardized PC score c, typically $c = \pm 3$.

3.4   Kendall's Spherical Coordinates

To visualize the shape distribution of triangles we use Kendall's spherical co-ordinate system. Using this coordinate system each triangle is mapped to one point on a sphere. The points on the southern hemisphere represent the reflection shape of those triangles on the northern hemisphere. Furthermore, the two poles of the sphere correspond to an equilateral triangle and its reflection shape, whereas the flat triangles are found in the regions close to the equator. In our case, the reflection shapes of the triangles have been removed after 3D point-based registration. Hence we need to consider only one hemisphere of Kendall's spherical coordinate system. Before constructing this coordinate system it is necessary to compute Kendall's coordinates $(u, v) \in \mathbb{R}^2$ for each triangle (for details we refer to [6]). The Kendall's coordinates can be converted into Kendall's spherical coordinates using the following formula:

$$x = \frac{1 - r^2}{2(1 + r^2)}, \quad y = \frac{u}{1 + r^2}, \quad z = \frac{v}{1 + r^2} \tag{4}$$

where $r^2 = u^2 + v^2$. Using (4) every triangle can be mapped to a point on the sphere.

Since we need to consider the shape distribution only on one hemisphere, we take advantage of the polar aspect of the Lambert-azimuthal equal-area projection to visualize our data. In this projection the north pole of the sphere is mapped to the center of one circle, whereas the equator is represented by the circle self.

3.5   Multidimensional Scaling (MDS)

To reconstruct the 3D structure of the BACs we apply multidimensional scaling (MDS). As input MDS uses a distance matrix. With this approach it is assumed that the shape variation of the BAC-triangles is relatively low. To establish the distance matrix in our application, we use the mean distances between each two BACs.

4   Experimental Results

For all datasets described in section 2 above we have applied 3D point-based rigid registration. As an example, Figs. 2a,b show the results of the registration for the real datasets AR1 and AR2. Figs. 2c,d visualize the datasets of the stable triangles and the random triangles. The registration removes the reflection shapes in the 2D plane. Therefore the transformed triangles can be evaluated using the complex Bingham distribution. Tab. 1 lists the values of $|\lambda_2|$ for all real datasets. The larger the value of $|\lambda_2|$, the lower is the shape variability of the triangles. For a comparison, we have also calculated the $|\lambda_2|$ value for the stable and random triangles yielding $|\lambda_s| = 162.52$ and $|\lambda_r| = 5.63$, respectively. Apparently, the real data are far from stable shapes. Except AR2 and AR4 all datasets are not random, since

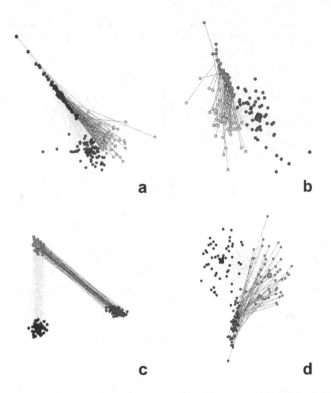

Fig. 2. Results after 3D point-based registration: Datasets AR1 (a), AR2 (b), stable triangles (c) and random triangles (d)

their $|\lambda_2|$ values are larger than $|\lambda_r|$. We have also applied the test of uniformity described in section 3.2 using a significance level of 1% yielding $\chi^2_{3;0.01} = 11.34$. The listed values for F in Tab. 1 reveal that all datasets except AR2 and AR4 are not uniformly distributed. This confirms the result using the complex Bingham distribution and the $|\lambda_2|$ values. Moreover, we can draw the same conclusion, if we use the heuristic criterion that the mean length of the triangles should be larger than three times the standard deviation of isotropic normally distributed vertices (which corresponds to a threshold value of $|\lambda_2| = 6.3$).

Table 1. Computed absolute values of λ_2 and result of the uniformity test for the real datasets. The last row lists the number of BAC-triangles for each dataset (sum: 457).

	Anti-ridge BACs					Ridge BACs				
	AR1	AR2	AR3	AR4	AR5	R1	R2	R3	R4	R5
$|\lambda_2|$	12.28	5.07	8.32	4.52	9.07	10.17	11.35	6.70	10.06	6.69
F	104.2	2.21	49.99	5.56	28.36	52.48	28.81	20.32	41.13	38.26
n	70	53	61	21	44	45	26	24	45	68

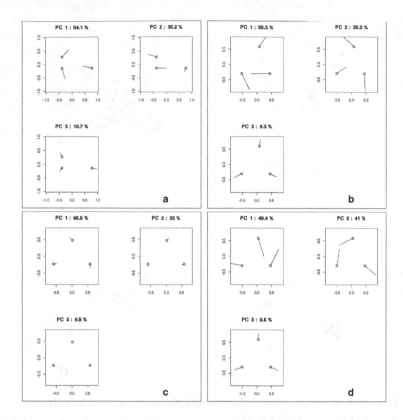

Fig. 3. The results of generalized Procrustes method and the corresponding percentage of variability captured by the i-th PC: Datasets AR1 (a), AR2 (b), stable triangles (c), and random triangles (d)

In Fig. 3 the results of the generalized Procrustes method are shown. Figs. 3a,b refer to the real datasets AR1 and AR2, and Figs. 3c,d to the stable and random triangles. The small circles represent the mean shape of the triangles (mean triangle). The vectors (circles attached to line segments) indicate the direction and magnitude of the variation along a certain principal component (PC) of the Procrustes residual. Generally the first PC captures the most dominant variation. In comparison to the random triangles (Fig. 3d), both real datasets have a larger dominant variation along the first PC compared to the other PCs. However, the other two PCs still have a relatively high variability, in particular, compared to the stable triangles. Note that the magnitude of the vectors for dataset AR1 is smaller than for dataset AR2. Analysing all 10 real datasets it turns out that all BACs-triangles possess high shape variability.

Fig. 4 illustrates the shape distribution of the BAC-triangles using Kendall's spherical coordinate system. The points for dataset AR1 are located primarily in one quarter of the large circle. In contrast, the points for the dataset AR2 are scattered randomly, which is similar to the random dataset. The points of

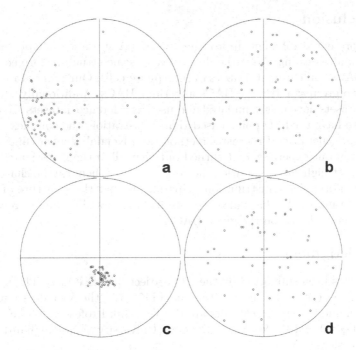

Fig. 4. Kendall's spherical coordinate system using Lambert's azimuthal equal-area projection. The north pole is represented by the center of the large circle. The results correspond to the datasets AR1 (a), AR2 (b), stable triangles (c), and random triangles (d).

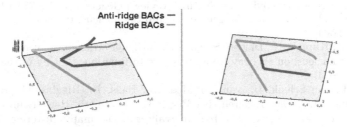

Fig. 5. Result of multidimensional scaling of anti-ridge and ridge BACs displayed from two different perspectives

the stable dataset aggregate into a small region as expected. These observations agree with the computed $|\lambda_2|$ values of the complex Bingham distribution.

Finally we show the feasibility of multidimensional scaling (MDS). Fig. 5 displays the polygon through all investigated five anti-ridge and five ridge BACs (based on mean distances between every two BACs as mentioned above). It turns out that the structure of the anti-ridge BACs coils more compactly than that of ridge BACs, which is what we expect. This is an interesting result, since we applied MDS although knew (based on the analysis above) that two of the ten dataset groups are randomly distributed.

5   Conclusion

We have presented different approaches based on statistical shape theory for analysing and assessing the variability of large-scale structure formed by gene regions (BACs) on chromatin fibers of interphase cells. Our real data is divided into the two groups: anti-ridge BACs and ridge BACs. Additionally, stable and random datasets have been simulated and used as reference datasets. To evaluate the data we have used 3D point-based rigid registration, the complex Bingham distribution, generalized Procrustes method, and Kendall's spherical coordinate system. From our experiments it turned out that all of the investigated dataset groups exhibit high shape variation, however, most of them can be characterized by a non-uniform shape distribution. This means that the structure of most of them is not random. We have also used multidimensional scaling to reconstruct the 3D structure based on the given BACs.

Acknowledgment

This work has been supported by the EU project 3DGENOME. The work benefited from the use of the Insight Toolkit (ITK) [10], the Visualization Toolkit (VTK) [11], and the statistical software R. We thank Professor John Kent (University of Leeds) and Professor Ian Dryden (University of Nottingham) for clarifying discussions.

References

1. Sproul, D., Gilbert, N., Bickmore, W.: The role of chromatin structure in regulating the expression of clustered genes. Nature Reviews Genetics 6, 775–781 (2005)
2. Verschure, P.J.: Positioning the genome within the nucleus. Biology of the Cell 96, 560–577 (2004)
3. Cremer, T., Küpper, K., Dietzel, S., Fakan, S.: Higher order chromatin architecture in the cell nucleus: on the way from structure to function. Biology of the Cell 96, 555–567 (2004)
4. Caron, H., van Schaik, B., van der Mee, M., Baas, F., Riggins, G., van Sluis, P., Hermus, M.C., van Asperen, R., Boon, K., Voute, P., Heisterkamp, S., van Kampen, A., Versteeg, R.: The human transcriptome map: clustering of highly expressed genes in chromosomal domains. Science 291, 1289–1292 (2001)
5. Kent, J.T.: The Complex Bingham Distribution and Shape Analysis. Journal of the Royal Statistical Society. Series B 56, 285–299 (1994)
6. Dryden, I., Mardia, K.: Statistical Shape Analysis. John Wiley & Sons, Chichester (1998)
7. Horn, B.: Closed-form solution of absolute orientation using unit quaternions. Journal of the Optical Society of America A 4, 629–642 (1987)
8. Mardia, K.V., Jupp, P.E.: Directional Statistics. John Wiley & Sons, Chichester (2000)
9. Gower, J.C.: Gerneralized procrustes analysis. Psychometrika 40, 33–50 (1975)
10. Ibanez, L., Schroeder, W., Ng, L., Cates, J.: The ITK Software Guide. Kitware, New York (2005)
11. Schroeder, W., Matin, K., Lorensen, B.: The Visualization Toolkit An Object-Oriented Approach To 3D Graphics, 3rd edn. Kitware, New York (2003)

Image Segmentation Using Histogram Fitting and Spatial Information

Da-Chuan Cheng[1], Xiaoyi Jiang[2], and Arno Schmidt-Trucksäss[3]

[1] Department of Radiological Technology, China Medical University, Taiwan
dccheng@mail.cmu.edu.tw
http://english.cmu.edu.tw
[2] Department of Mathematics and Computer Science, University of Münster, Germany
xjiang@math.uni-muenster.de
http://cvpr.uni-muenster.de
[3] Lehrstuhl und Poliklinik für Präventive und Rehabilitative Sportmedizin, Technische Universität München, Germany
schmidt-trucksaess@sport.med.tum.de
http://www.sport.med.tum.de

Abstract. In this paper, we introduce a novel unsupervised segmentation method using a histogram fitting method to find out the optimal histogram clustering based on multi Gaussian models. The fitting problem is performed via the trust region reflective Newton method to minimize a predefined cost function. The histogram clustering is the global information describing the probability of a given gray value belonging to a category. Together with the consideration of the spatial information, the image segmentation is performed. We demonstrate some applications on medical images such as brain CT and MRI.

Keywords: histogram clustering, curve fitting, trust-region method, image segmentation.

1 Introduction

Image segmentation is a basic step in many image processing techniques. Especially, it is a pre-processing and low-level process in the computer vision field. This is an old but important topic in many medical applications. The general goal of segmentation is to group the image primitives having similarity together. The image primitives can be low-dimensional such as pixel gray level, texture, or they can be hyper-dimensional such as a composition of many useful features.

There are many histogram based image segmentation articles over the past decade. It is difficult to categorize them because the amount of related studies is too large. We however divide them into two groups: parametric [1,2] and nonparametric method [3]. Normally, the parametric method can handle the unsupervised image segmentation problem because they use some models to represent the characteristics of objects or background. On the contrary, the nonparametric methods such as [4] usually need some information as training data or references to help performing the segmentation task.

P. Perner and O. Salvetti (Eds.): MDA 2006/2007, LNAI 4826, pp. 47–57, 2007.

Also, the case-based reasoning methods [5] for image segmentation have been developed that adjust the model parameters for the processed image by retrieving past successfully solved cases from a case base and that can incrementally learn new case solutions to improve the segmentation quality.

In the case of unsupervised clustering, image segmentation is performed by setting a model describing the nature of the feature distribution and finding out the optimal model parameters without any prior knowledge (training data). The prerequisite is that the model should be general and available for this kind of images since this is the only framework information. Different from the supervised clustering, the feature distribution for unsupervised clustering is mostly assumed to be compact and hyperellipsoidal [6,7]. This is because we have no priori knowledge on the feature distribution. A frequently used model is normal distribution [8]. Other models used in computer vision can be found in [9]. However, the distribution can also be assumed to be non-hyperellipsoidal but continuous and locally linear such as locally linear embedding method [10,11].

On the other aspect, the feature selection is also important. A good feature selection can well represent the objects so that the following clustering then becomes easier. Since we do not know the distribution in advance, it is easier to solve the problem by supervised clustering methods. With the help of training data, the feature distribution can be sampled and the boundaries can be found, for instance by support vector machine [12,13]. However, based on a reasonable assumption that spatially close pixels are likely to belong to the same tissue type [14], the segmentation can be better done with the consideration on the features (either local or global information) together with the spatial information. We human beings recognize objects also based on the features and the spatial information. We therefore have to consider the coordinate relationship. Simultaneously, the similar object can also appear at another place so that we need global information such as colors or textures. The segmentation can be well done only when these two information sources are simultaneously considered. For example, in [2] the image segmentation is performed via annealing maximum a posteriori estimation to compute the optimal histogram clustering solutions. With a suitable feature selection method, a simple histogram clustering method can segment texture images as well.

In this paper, we introduce a method using unsupervised histogram clustering via a curve fitting process based on multi Gaussian models. The histogram is an invariant feature available for medical image processing. Simultaneously, we take the spatial information into consideration to reduce the noise effect and to connect the thin object which is disconnected by noises.

The paper is organized as follows: We propose a model using a curve fitting technique to solve the histogram clustering problem in section 2. Then a segmentation approach is introduced using the spatial information as auxiliary knowledge in grouping objects in section 3. In section 4 we demonstrate examples of segmentation on CT and MRI brain images to show the performance of our algorithm. The final discussion and conclusion are issued afterwards.

2   Histogram Fitting Model

2.1   Basis Function

Curve fitting is a process of finding a curve which matches a series of data points and possibly other constraints based on a predefined model. Normally we have to choose a model composed of many basis functions to describe the curve to be fitted. For example, the polynomial function, the sinusoidal function, or the exponential function can be used as basis functions to form a model. In this paper, we use Gaussian-like function as our basis function. It is based on the assumption that every object (tissue type) in the medical image has a Gaussian-like distribution in the histogram. The clustering task is transformed to a fitting problem, which fits the histogram by a given number of model distributions. The potential of the feature belonging to some cluster is then determined by the probability of the model distribution with the consideration of the local spatial information. We will discuss it more in details in section 3.

Therefore, three parameters describe the i-th distribution as follows:

$$p_i(x; a_i, b_i, c_i) = p_i(x) \; = \; a_i \phi_i(x) \; = \; a_i \exp(-((x - b_i)/c_i)^2), \qquad (1)$$

$$\text{with the constraints: } a_i > 0, \; b_i \geq 0 \text{ and } c_i \neq 0$$

where a_i is the intensity (or weighting), b_i is the translation on the gray level and c_i is the standard deviation. The function is slightly different from the normal distribution since the integration of this function does not have to be 1. In order to have a short form, we use $p_i(x)$ instead of the complete form $p(x; a_i, b_i, c_i)$. The parameter set $v_i = (a_i, b_i, c_i)$ is omitted in the following paragraphs.

The cluster number determination is an important aspect, however, it is not within the scope of this study. In treating the medical images such as CT or MRI, we normally know the cluster number in advance. Assuming there are k clusters, our model to fit the histogram can then be formulated as follows:

$$p(x) = \begin{cases} p_1(x) & 0 \leq x < th_1 \\ p_i(x) & th_i \leq x < th_{i+1}, \; 1 < i < k \\ p_k(x) & th_k \leq x \leq 255 \end{cases} \qquad (2)$$

where th_i denotes the ith threshold, which is the cross point of two Gaussian-like functions. The thresholds are determined directly by checking their relationship. The technical details are given in the appendix. Note that b_i's denoting the centers of the Gaussian-like distributions are not necessary to be in an ascending order (see Fig. 1.) The reason is that the mean center can be anywhere for representing an arbitrary distribution on the histogram. The basis function is ordered by index i which is used to determine the range of fitting period.

Once the model is determined by $3k$ parameters, the cost function of the fitting problem is then easily defined by sum of the least squares:

$$v_{opt} \; = \; \arg\min_v \sum_x (h(x) - p(x))^2 \qquad (3)$$

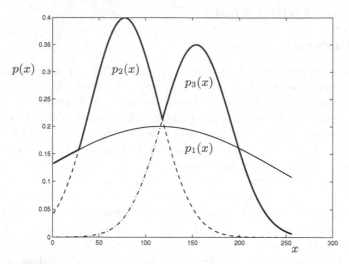

Fig. 1. The thick solid curve is the model curve which is composed of three Gaussian-like functions: $(b_1, b_2, b_3) = (115, 77, 154)$

where $v = \{v_i\}$ is the parameter set of the model $p(x)$ and h(x) is the curve to be fitted. The trust-region reflective Newton method [15,16] is applied as an iterative optimization tool to find out the optimal parameter set v; see Fig. 2 for an example of this fitting method on a real histogram.

2.2   Scaling Problem

The scaling is very important in the fitting and optimization process including its convergence efficiency. In a medical image it is very often that the background contains large area. Therefore, it results in a huge peak on the histogram and it causes problems in fitting. Here we use a power instead of the original histogram to scale the ordinate. The transformation is defined as follows:

$$h(x) = h'^n(x), \quad n = \frac{1}{\lfloor \log(N) \rfloor} \tag{4}$$

where N is the pixel number of the image. Here the power is an elementwise operation. Afterwards, $h(x)$ is normalized between $[0, 1]$ and then can be used in Eq.(3). The scaling on the abscissa is linearly transformed from $[0, 255]$ to $[0, 10]$.

2.3   Initialization and Optimization

Normally, the initialization is an important aspect in the iterative optimization process. This is because the minimization might fall into local minimum which is not the optimal solution. The initializations of b_i are uniformly distributed on the abscissa according to cluster number k. The initial values of a_i are the samples on $h(b_i)$ and the initial values of $c_i = 1$. The lower and upper bounds are set and the trust-region reflective Newton method is applied to iteratively solve the minimization problem. Here a total of $3k$ parameters have to be solved iteratively.

3   Segmentation

Image segmentation is a process to distinguish objects from background. The applications on medical images are to distinguish different tissues. Traditionally, four popular categories are classified: threshold techniques, edge-based method, region-based methods, and connectivity-preserving relaxation methods. The medical images are quite different from popular images obtained from cameras. They are specific and there are anatomical models or knowledge to increase the accuracy of segmentation techniques. In this study, we focus on medical images such as CT and MRI. In this kind of images, the tissue segmentation can be done via a thresholding technique and with the considerations on spatial information.

After the histogram clustering is done with the histogram fitting models described in section 2, all thresholds can be obtained. We then establish a label matrix B of the same size as the raw image R. In the label matrix all pixels are represented by label index, i.e. $[1, \cdots, C]$. In matrix B we found many isolated pixels like noises. Normally, the tissues are connected without isolated pixels. In order to avoid the disconnections, the spatial information has to be considered.

To ease the computation, total C matrices are set:

$$M_l(x, y) = \begin{cases} 1 & \text{if } B(x, y) = l \\ 0 & \text{otherwise.} \end{cases} \quad \text{for all } (x, y) \text{ and } 1 \leq l \leq C.$$

We examine every point (x, y) on B and count the number of point having the same label, i.e. $l = B(x, y)$, around this point on $M_l(x, y)$ in a neighborhood region Ω.

$$N_l = \sum_{(x+\delta_1, y+\delta_2) \in \Omega} M_l(x + \delta_1, y + \delta_2), \tag{5}$$

where Ω is the area of the size 7×7 and centered at (x, y). If $N_l > 42$, then class l has major neighborhood points having the same labels. We then set the resulting label matrix $\hat{B}(x, y) = l$. Otherwise, we have to consider if it is a thin object. The value 42 is chosen because the mask width is $49 - 42 = 7$ so that it is possible to detect thin objects.

For this purpose some masks are defined:

$$mask_1 = \begin{bmatrix} 0 & 0 & 0 & v_1 & 0 & 0 & 0 \\ 0 & 0 & 0 & v_2 & 0 & 0 & 0 \\ 0 & 0 & 0 & v_3 & 0 & 0 & 0 \\ 0 & 0 & 0 & v_4 & 0 & 0 & 0 \\ 0 & 0 & 0 & v_3 & 0 & 0 & 0 \\ 0 & 0 & 0 & v_2 & 0 & 0 & 0 \\ 0 & 0 & 0 & v_1 & 0 & 0 & 0 \end{bmatrix}$$

where the factors v_i denote the weightings at the respective positions. The $mask_1$ considers the thin structure such as vertical lines. Moreover, a horizontal and two diagonal structure matrices are constructed for detecting thin lines of different directions, in which v_4 is always at the center. We denote them as $mask_2$, $mask_3$ and $mask_4$.

For every point on B whose corresponding N is less than 43, we then examine if it is a thin object. The weightings in $mask_1$ are given as $[v_1, v_2, v_3, v_4] = [0.7\ 0.8\ 0.9\ 1]$. The points near the center get larger weightings to emphasize the connectivity. All labels appear in current Ω centered at $\boldsymbol{x} = (x, y)$ are saved in a vector $l_{\boldsymbol{x}}$.

$$w(l, i) = \sum_{(x+\delta_1, y+\delta_2) \in \Omega} M_l(x + \delta_1, y + \delta_2) \cdot mask_i(\delta_1, \delta_2)$$

$$\text{for all } l \in l_{\boldsymbol{x}} \text{ and } 1 \leq i \leq 4.$$

The larger value in w denotes the larger connectivity and homogeneity of this thin object.

$$l_{max} = \max_l w(l, i)$$

Thus we assign $\widehat{B}(x, y) = l_{max}$.

4   Results and Discussion

We have tested some CT brain images with the novel algorithm. Due to page limitation only one example is shown here. Figure 2 demonstrates one of the tests using our fitting model.[1] Fig.2(c) is the brain CT image. Its histogram and the fitting curve are shown in Fig.2(a). The total cluster number is five and the basis functions are shown in Fig.2(b) superimposed on the histogram. The different clusters are shown by five gray levels in Fig.2(d).

Figure 3 demonstrates a result that our method is applied on the CT images having a tumor. There are many images in this sequence and the other results are similar. This result shows the ability of segmenting tumors in CT images.

As an example we also demonstrate our algorithm on MRI brain images. Figure 4 shows automated clustering and segmentation results.

An important aspect is the way of initialization. We use in this study uniformly distributed seeds for the parameter b_i. This way is not robust. However, we also found that once the initializations for b_i are approaching to the "correct" positions, then the initializations for parameters c_i are not important. They can easily converge to the global/local minimum.

The method is unsupervised because it does not need any priori knowledge on the probability distribution of the feature histogram. The determination of cluster number is given manually which is not within the scope of this study.

---

[1] A PDF file of this paper with color figures can be downloaded from http://wwwmath. uni-muenster.de/u/xjiang/papers/MDA2007.pdf

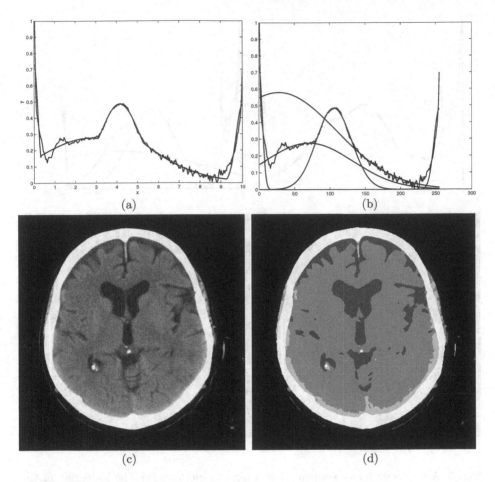

Fig. 2. A CT brain image segmentation using our method: (a) The histogram of the image in (c). The fitting curve (the smoother curve) is superimposed on the histogram. (b) All basis functions and the histogram. (c) Brain CT image. (d) The segmentation result in five gray levels. Cluster number $k = 5$.

The program is written on the Matlab platform [16]. The optimization process is the trust region reflective Newton. The computation time for curve fitting is around 2 sec on a PC with a 1.8 GHz CPU. The segmentation takes longer and it depends on the image size. For an image of the size 460×480 it takes about 8 to 10 sec.

5   Conclusion

We have developed a novel algorithm using the fitting technique to solve the histogram clustering problem and consider the spatial information as the auxiliary knowledge in grouping objects. This algorithm is simple to be implemented.

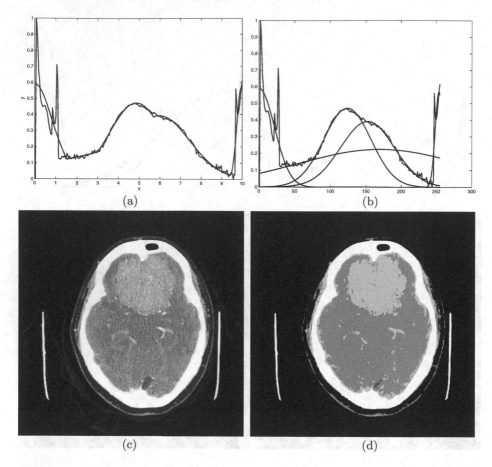

Fig. 3. A CT brain image segmentation using our method: (a) The histogram of the image in (c). The fitting curve (the smoother curve) is superimposed on the histogram. (b) All basis functions and the histogram. (c) Brain CT image having a tumor. (d) The segmentation result in five gray levels. Cluster number $k = 5$.

Some experiments are made on CT and MRI brain images and the results are convincing to be useful in segmenting such as bone, brain tissues, CSF, and hematoma. This measuring system is useful for clinical quantification studies. The future goals of this study are to automatically determine the cluster number and to give a robust initialization for the iterative optimization process.

Acknowledgement

The authors would like to thank Dr. Yu-Chien Lo for providing the tumor CT image in figure 3. The financial support is from China Medical University, Taiwan.

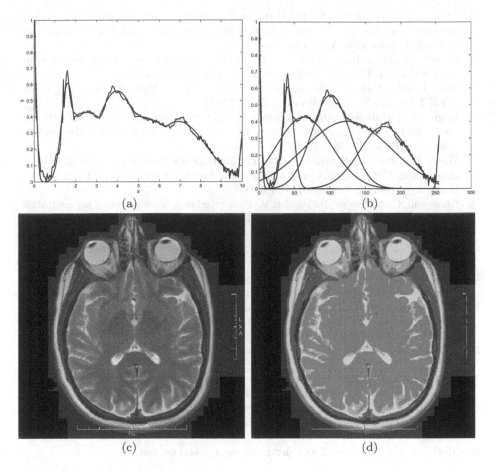

Fig. 4. A MRI brain image segmentation using our method: (a) The histogram of the image in (c). The fitting curve (the smoother curve) is superimposed on the histogram. (b) All basis functions and the histogram. (c) Brain MRI image. (d) The segmentation result in seven gray levels. Cluster number $k = 7$.

References

1. Hermes, L., Zoeller, T., Buhmann, J.M.: Parametric distributional clustering for image segmentation. In: Tistarelli, M., Bigun, J., Jain, A.K. (eds.) ECCV 2002. LNCS, vol. 2359, pp. 577–591. Springer, Heidelberg (2002)
2. Puzicha, J., Buhmann, J.M., Hofmann, T.: Histogram clustering for unsupervised image segmentation. In: Proc. of IEEE Conf. on Computer Vision and Pattern Recognition (CVPR), vol. 2, pp. 2602–2608. IEEE Computer Society Press, Los Alamitos (1999)
3. Kim, J., Fisher III, J.W., Yezzi, A., Cetin, M., Willsky, A.: A nonparametric statistical method for image segmentation using information theory and curvature evoluation. IEEE Trans. on Image Processing 14(10), 1486–1502 (2005)

4. Chan, T., Esedoglu, S., Ni, K.: Histogram based segmentation using wasserstein distances. In: The 1st International Conference on Scale Space and Variational Methods in Computer Vision (in press, 2007)
5. Perner, P.: An architecture for a CBR image segmentation system. Engineering Applications of Artificial Intelligence 12(6), 749–759 (1999)
6. Mao, J., Jain, A.K.: A self-organizing network for hyperellipsoidal clustering (hec). IEEE Trans. on Neural Network 7, 16–29 (1996)
7. Frigui, H., Krishnapuram, R.: A robust competitive clustering algorithm with applications in computer vision. IEEE Trans. on Pattern Analysis and Machine Intelligence 21(5), 450–465 (1999)
8. Wang, Y., Adah, T., Kung, S., Szabo, Z.: Quantification and segmentation of brain tissues from MR images: A probabilistic neural network approach. IEEE Trans. on Image Processing 7(8), 1165–1181 (1998)
9. Chesnaud, C., Refregier, P., Boulet, V.: Statistical region snake-based segmentation adapted to different physical noise models. IEEE Trans. on Pattern Analysis and Machine Intelligence 21(11), 1145–1157 (1999)
10. de Silva, V., Tenenbaum, J.B.: Global versus local methods in nonlinear dimensionality reduction. In: Proc. NIPS, vol. 15, pp. 721–728 (2003)
11. Saul, L.K., Roweis, S.T.: Think globally, fit locally: Unsupervised learning of low dimensional manifolds. Journal of Machine Learning Research 4, 119–155 (2003)
12. Boser, B., Guyon, I., Vapnik, V.: A training algorithm for optimal margin classifiers. In: Haussler, D. (ed.) 5th Annual ACM Workshop on COLT, pp. 144–152. ACM Press, Pittsburg, PA, USA (1992)
13. Schölkopf, B., Smola, A.J.: Learning with Kernels. MIT Press, Cambridge, MA (2002)
14. Bouman, C., Liu, B.: Multiple resolution segmentation of texture images. IEEE Trans. on Pattern Analysis and Machine Intelligence 13, 99–113 (1991)
15. Coleman, T.F., Li, Y.: A reflective newton method for minimizing a quadratic function subject to boundson some of the variables. Technical Report TR92-1315 (1992)
16. Matlab:The mathworks (2006), http://www.mathworks.com

Appendix

The cross points of two Gaussian-like functions defined as $p_1(\cdot)$ and $p_2(\cdot)$ can be obtained directly as follows. Let (a_1, b_1, c_1) and (a_2, b_2, c_2) denote the parameter sets of the two functions. In order to obtain the cross points, we have to solve the equation:

$$a_1 \exp\left(\frac{-(x - b_1)^2}{c_1^2}\right) = a_2 \exp\left(\frac{-(x - b_2)^2}{c_2^2}\right)$$

where $a_i > 0$, $b_i \geq 0$, and $c_i \neq 0$. Therefore, the equation can be solved using nature log functions on both sides:

$$\ln a_1 - \frac{(x - b_1)^2}{c_1^2} = \ln a_2 - \frac{(x - b_2)^2}{c_2^2}$$

$$\Rightarrow \underbrace{(c_2^2 - c_1^2)}_{A} x^2 + \underbrace{(2b_2 c_1^2 - 2b_1 c_2^2)}_{B} x + \underbrace{(c_2^2 b_1^2 - c_1^2 b_2^2 - c_1^2 c_2^2 (\ln a_1 - \ln a_2))}_{C} = 0$$

This is a quadratic equation and if $B^2 - 4AC > 0$, then it has two real solutions. Assuming q_1 and q_2 are the two cross points and $q_1 < q_2$. We choose q_1 as the threshold point if it satisfies:

$$p_1(q_1 - \delta) > p_2(q_1 - \delta),$$

otherwise we choose q_2 as the threshold point if it satisfies:

$$p_1(q_2 + \delta) < p_2(q_2 + \delta),$$

where $\delta > 0$ is a small constant. If $c_1 = c_2$ and $b_1 \neq b_2$, then it reduces to one solution and this solution is the threshold point. The threshold values are determined via checking the relationship between these two functions. Theoretically, it is possible to get none real solution. In this case, we give a random value vector instead of $p(x)$. This is a technique to increase the error in the optimization process.

Statistical Analysis of Microspectroscopy Signals for Algae Classification and Phylogenetic Comparison

Anna Tonazzini, Primo Coltelli, and Paolo Gualtieri*

Istituto di Scienza e Tecnologie dell'Informazione - CNR
Via G. Moruzzi, 1, I-56124 PISA, Italy
anna.tonazzini@isti.cnr.it, primo.coltelli@isti.cnr.it
Istituto di Biofisica - CNR
Via G. Moruzzi, 1, I-56124 PISA, Italy
paolo.gualtieri@pi.ibf.cnr.it

Abstract. We performed microspectroscopic evaluation of the pigment composition of the photosynthetic compartments of algae belonging to different taxonomic divisions and higher plants. In [11], a supervised Gaussian bands decompositions was performed for the pigment spectra, the algae spectrum was modelled as the linear mixture, with unknown coefficients, of the pigment spectra, and a user-guided fitting algorithm was employed. The method provided a reliable discrimination among chlorophylls a, b and c, phycobiliproteins and carotenoids. Comparative analysis of absorption spectra highlighted the evolutionary grouping of the algae into three main lineages in accordance with the most recent endosymbiotic theories. In this paper, we adopt an unsupervised statistical estimation approach to automatically perform both Gaussian bands decomposition of the pigments and algae fitting. In a fully Bayesian setting, we propose estimating both the algae mixture coefficients and the parameters of the pigment spectra decomposition, on the basis of the alga spectrum alone. As a priori information to stabilize this highly underdetermined problem, templates for the pigment spectra are assumed to be available, though, due to their measurements outside the protein moiety, they differ in shape from the real spectra of the pigments present in nature by unknown, slight displacements and contraction/dilatation factors. We propose a classification system subdivided into two phases. In the first, the learning phase, the parameters of the Gaussians decomposition and the shape factors are estimated. In the second phase, the classification phase, the now known real spectra of the pigments are used as a base set to fit any other spectrum of algae. The unsupervised method provided results comparable to those of the previous, supervised method.

---

* This work has been partially supported by the European project Network of Excellence MUSCLE FP6-507752 (Multimedia Understanding through Semantics, Computation and Learning).

P. Perner and O. Salvetti (Eds.): MDA 2006/2007, LNAI 4826, pp. 58–68, 2007.

1   Introduction

The term algae has no formal taxonomic standing, nevertheless it is routinely used to indicate a polyphyletic, non-cohesive and artificial assemblage, of O_2-evolving, photosynthetic organisms. No easily definable classification system acceptable to all exists for algae, since taxonomy is under constant and rapid revision at all levels following every day new genetic and ultrastructural evidence. Keeping in mind that the polyphyletic nature of the algal group is somewhat inconsistent with traditional taxonomic groupings, though they are still useful to define the general character and level of organization, and aware of the fact that taxonomic opinion may change as information accumulates, we have adopted a scheme of classification mainly based on that of Van Den Hoek *et al.* [1]. According to the most recent theories, different evolutionary lineages can be recognized within the algal world [2]. Three major eukaryotic photosynthetic groups have descended from a common prokaryotic ancestor through an endosymbiotic event. The three lineages of primary plastids were found in the Glaucophyta, in the green algae and plants, and in the red algae. Photosynthetic compartments contain the pigments for absorbing light and channeling the energy of the excited pigment molecules into a series of photochemical and enzymatic reactions. All those pigments are organized in supra-molecular structures of pigment-protein complexes embedded in the membrane of sac-like flat compressed vesicles, the thylakoids. The pigments present in algal cells (i.e. different type(s) of chlorophylls, different type(s) of carotenoids, and different type(s) of phycobiliproteins) provide a convenient paradigm to explain evolutionary development involving endosymbiotic acquisition of photosynthetic cellular organelles. The absorption spectra measured *in vivo* on region of the photosynthetic compartment, i.e. microspectrophotometry, can give us very precise and accurate information about the spectral range in which pigment molecules organized in the thylakoid membranes capture photons in their natural environment [3]. Since each pigment possesses its own distinctive absorption spectrum in the visible range, and their combination constitutes the unique absorption spectrum of thylakoid compartment of the alga, the high quality absorption spectra obtained by means of microspectroscopy can be used to discriminate effectively the pigment molecules that contribute to the whole spectrum. The use of microspectrophotometry can expand the perception of taxonomists, who identify algae in relation to natural pigmentation to supplement classification based on morphology. As consequence it is possible to predict the presence of a specific pigment in an alga, to give an unknown alga a plausible taxonomic framing, and to contribute to support the phylogenetic tree of the endosymbiotic events. Over the last 30 years there have been considerable efforts to identify the components of algal absorption using mathematical techniques [4][5] [6][7][8][9]. These methods produced good results in the identification of spectral peaks of light absorbing pigments.

In this paper, we propose a method to classify algae by identifying the major absorption peaks in the eleven algal divisions *sensu* Van der Hoeck [1]. First, the alga spectrum is measured by using microspectroscopy, and then modelled as the linear mixture, with unknown coefficients, of a set of unknown pigment spectra. We propose analyzing the alga spectrum by means of a statistical blind estimation approach, in order to detect the pigments and their relative weights (the mixture coefficients). Adopting a fully Bayesian blind estimation method, both the mixture coefficients and the pigment spectra should be estimated on the basis of the alga spectrum alone. A priori, biologically grounded, knowledge is however exploited to stabilize the highly underdetermined problem. In particular, we assume that each pigment spectrum is, in turn, represented by a mixture of Gaussians, whose weights and parameters must be estimated as well. Templates for the pigment spectra are also available, though, due to their measurements outside the protein moiety, they differ in shape from the real spectra of the pigments present in nature by unknown, slight displacements and contraction/dilatation factors. We propose a classification system subdivided into two phases. In the first, the learning phase, a given, possibly large, set of spectra of algae, each containing an unknown subset of the pigments for which a template is available, are fitted using the pigments themselves as bases, in order to estimate the parameters of the Gaussians mixtures and the shape factors. Once the system has learned the parameters, in the second phase, the classification phase, the real spectra of the pigments are assumed known and used as a base set to fit any other spectrum of algae whose pigments are unknown. Since in nature pigments are of the order of hundreds, for each new alga a "residual", unknown pigment, still in the form of Gaussian mixture, can be eventually estimated together with the weights of the pigments given.

2   Materials and Methods

2.1   Algae Cultures

All the cultures were grown in chemically defined media as described in Barsanti et al.[11]

2.2   Absorption Microscopy

Absorption spectra in the visible range, from 400 nm to 700 nm, were measured *in vivo* on photosynthetic compartments (thylakoid membranes, or chloroplasts) of single cells belonging to the different algal divisions. The apparatus to perform *in vivo* measurements was previously described in details [10]. All the absorption spectra were recorded from 400 nm to 700 nm, with a step size of 0.5 nm and scan speed of 100 $nm * sec^{-1}$. For each wavelength, 10,000 values of optical density were averaged. The resolution achieved with this step size and wavelength width was sufficient to distinguish between the major pigment classes present in the photosynthetic compartments.

3   Formulation of the Problem

We formulate the problem of classifying algae on the basis of the pigments they contain as the problem of estimating both the coefficients and the pigment bases of the linear mixtures which models the spectrum of the alga measured by microspectroscopy. Such a problem falls in the class of problem that can be tackled by means of Blind Source Separation (BSS) techniques. BSS became an active research topic in statistical signal processing in the last decade [15][12][18], with applications in audio processing, removal of underlying artifact components of brain activity from EEGs, search for hidden factors in parallel financial series, and, more recently, in image processing and computer vision [16], for feature extraction or noise removal, for separation of components in astrophysical microwave maps [19] and for document restoration [20]. BSS consists of separating a set of unknown signals from a set of mixtures of them, when no full knowledge is available about the mixing operator. Assuming that the signals are mutually independent and the data are noiseless, separation techniques based on independent component analysis (ICA) have been proven to perform satisfactorily in many applications. Independence is however a strict requirements which cannot be fulfilled in most practical cases. The literature regarding approaches to BSS able to cope with noisy data, cross-correlation among the sources and autocorrelation inside the single sources, different numbers of sources and data signals, convolutive or nonlinear mixtures, is by now very rich [14][17][13]. Among the others, the Bayesian estimation setup seems to be very promising to solve various instances of BSS, since it offers a natural and flexible way to approach the integrated solution of two or more problems, and to account for any prior knowledge we may have about a problem [21].

According to the BSS formalism, the generic data model we consider is:

$$\mathbf{xa}(\lambda) = A\mathbf{s}(\lambda) + \mathbf{n}(\lambda) \qquad \lambda = 1, 2, ..., \Lambda \tag{1}$$

where $\mathbf{xa}(\lambda)$ is the vector of the measurements, the alga spectra, $\mathbf{s}(\lambda)$ is the column vector of the unknown sources, i.e. the pigment spectra, and $\mathbf{n}(\lambda)$ is the noise or measurement error vector, at wavelength λ, and A is the unknown mixing matrix, assumed location-independent. We assume different numbers Na and Np of measured and source signals, with $Na \geq Np$, so that A is an $Na \times Np$ matrix. In a fully Bayesian approach, both A and $\mathbf{s}$ are assumed as independent unknowns, and are assigned with prior distributions $P(A)$ and $P(\mathbf{s})$, respectively. Then, at least in principle, A and $\mathbf{s}$ can be simultaneously estimated, according to the Maximum A Posteriori (MAP) estimation criterion, by maximizing the posterior distribution $P(\mathbf{s}, A|\mathbf{xa})$:

$$P(\mathbf{s}, A|\mathbf{xa}) \propto P(\mathbf{xa}|\mathbf{s}, A)P(\mathbf{s})P(A) \tag{2}$$

or, equivalently, by minimizing the negative logPosterior, or energy function, $E(\mathbf{s}, A|\mathbf{xa})$:

$$E(\mathbf{s}, A|\mathbf{xa}) = -logP(\mathbf{xa}|\mathbf{s}, A) - logP(\mathbf{s}) - logP(A) \tag{3}$$

where $\mathbf{xa} = (\mathbf{xa}(1), ..., \mathbf{xa}(\Lambda))$, $\mathbf{s} = (\mathbf{s}(1), ..., \mathbf{s}(\Lambda))$, and $P(\mathbf{xa}|\mathbf{s}, A)$ is the likelihood, i.e. the noise distribution. Since the noise is basically due to the measurement process, it is reasonable to consider it as a white, Gaussian and stationary process with zero mean. The choice of proper $P(A)$ and $P(\mathbf{s})$ is fundamental to restrict the set of solutions associated to the likelihood part of eq. 3.

According to the above formulation, for each set of algae to be classified, only a number of pigments smaller or equal to the number of measured algae can be estimated. This presents two orders of inconvenient. First, we know that in nature pigments are of the order of hundreds; second, some algae share the same pigments. Thus, re-estimating them each time a new set of algae is measured becomes uneconomic. We then propose to devise a system subdivided into two phases: a learning phase and a classification phase. In the learning phase, the system is provided with a large set of algae belonging to the eleven divisions identified by the classification scheme of Van Den Hoek *et al.* (1995) [1]. The pigments of these algae belong to a set of predefined, though not fully known, pigments, characteristic of those divisions. The above formulation of the problem allows the estimation of these pigments along with the algae mixing coefficients. In the classification phase, any alga can be given as input to the system and, exploiting the now fully known pigments, only the weights of the mixture need to be estimated. The principle, in both cases, is the minimization of eq. 3, but with respect to both A and $\mathbf{s}$ in the learning phase, and restricted to the only A in the classification phase. For a better fitting, however, in the classification phase the problem is augmented with a new, unknown "residual" pigment, still in the form of Gaussian mixture, to be estimated together with the weights of the pigments given. These two phases will be described in details in the two subsequent sections, where the specific form adopted for $P(A)$ and $P(\mathbf{s})$ will be also detailed.

4   The Learning Phase

In the learning phase, we assume the availability of a set $\mathbf{xa}$ of Na real spectra of algae, each containing an unknown subset of a fixed set $\mathbf{s}$ of Np pigments. As already said, the algae spectra are assumed to be linear mixtures, with unknown coefficients A_{ij}, $i = 1, ..., Na$, $j = 1, ..., Np$ of the pigment spectra. Thus, in the assumption of a white, Gaussian noise with zero mean, the logarithm of the likelihood $log(P(\mathbf{xa}|\mathbf{s}, A)$ is given by:

$$log(P(\mathbf{xa}|\mathbf{s}, A) = - \sum_{i=1}^{Na} \sum_{\lambda=1}^{\Lambda} \left(xa_i(\lambda) - \sum_{j=1}^{Np} A_{ij} s_j(\lambda) \right)^2 \tag{4}$$

where Λ represents the number of sampling data points in the range $[400nm, 700nm]$. The set of fixed pigments needs not to be fully known, but some information about them should be available in order to define the prior $P(\mathbf{s})$.

In this respect, the spectra of the pigments are assumed to be well represented by mixtures of Gaussians, whose number NG_j, parameters $(\lambda_{jk}, \sigma_{jk})$ and weights w_{jk}, $j = 1, ..., Np$, $k = 1, ..., NG_j$ are however unknown. The Gaussian model constitutes a first good approximation usually adopted in the literature. This spectral decomposition could be equally performed using Lorentzian or Voigt profiles (convolution of Lorentzian and Gaussian). However, we used Gaussian fitting because the signal measured by a spectrophotometer (i.e. the spectrum) is transformed by the instrument into a Gaussian curve.

Furthermore, we assume the availability of a set $\mathbf{xp}$ of Np templates for the spectra of these pigments. Due to the method adopted to construct the pigment templates, it is expected that slight factors of displacement $\mathbf{d}_j$ and contraction/dilatation $\mathbf{c}_j$, $j = 1, ..., Np$, will affect each pigment template with respect to the real pigment spectrum, so that these templates cannot be directly used as bases for algae fitting.

According to the Bayesian formalism described in the previous section, the Gaussian mixture modelling together with the pigment models and templates represent the a priori knowledge we have on the source pigments, and can be expressed through $P(\mathbf{s})$ in the following way:

$$logP(\mathbf{s}) = -\sum_{j=1}^{Np}\sum_{\lambda=1}^{\Lambda} (xp_j(\lambda) - s_j(\lambda; \mathbf{d}_j, \mathbf{c}_j))^2 \tag{5}$$

where

$$s_j(\lambda) = \sum_{k=1}^{NG_j} w_{jk} exp\left(\frac{-(\lambda - \lambda_{jk})^2}{2\sigma_{jk}^2}\right) \quad \forall \lambda, j = 1, ..., Np \tag{6}$$

and

$$s_j(\lambda; \mathbf{d}_j, \mathbf{c}_j) = \sum_{k=1}^{NG_j} w_{jk} exp\left(\frac{-(\lambda - (\lambda_{jk} + d_{jk}))^2}{2(\sigma_{jk} + c_{jk})^2}\right) \quad \forall \lambda, j = 1, ..., Np \tag{7}$$

It is now clear that minimizing the energy function in eq. 3 with respect to A and $\mathbf{s}$ is equivalent to minimize it with respect to A and all the parameters that define the Gaussian mixtures in eqs. 6-7. By calling Ω the set of these parameters, the overall energy function to be minimized in the learning phase is thus given by:

$$E_l(\Omega) = \sum_{j=1}^{Np}\sum_{\lambda=1}^{\Lambda} (xp_j(\lambda) - s_j(\lambda; \mathbf{d}_j, \mathbf{c}_j))^2 + \beta \sum_{i=1}^{Na}\sum_{\lambda=1}^{\Lambda}\left(xa_i(\lambda) - \sum_{j=1}^{Np} A_{ij} s_j(\lambda)\right)^2 - logP(\Omega) \tag{8}$$

in view of eqs. 6-7, and where $P(\Omega)$ is a prior on Ω, expressing all other information, such as bounds, positivity and so on, we may have on the parameters, and including $P(A)$. Parameter β is a weight balancing the fidelity of the two data sets, algae and pigment templates.

When attempting to fit a distribution or signal with a Gaussian mixture, the choice or estimation of the number of Gaussian components is a very critical issue. For an arbitrary number of Gaussians, the fitting algorithm will always tend to use all of them. This, of course, can give rise to solutions that are biologically unfeasible. Thus, the number of Gaussians for a given pigment should be the minimum compatible with biological plausibility and knowledge. An attempt to automatically determine the optimal number should exploit the above information under the form of suitable constraints to be added to the energy function. However, in this first application of our approach, we assume the exact knowledge of the number of Gaussian components for each real pigment spectrum, so that the set of parameters to be estimated results in:

$$\Omega = \{A, \mathbf{w}, \lambda, \sigma, \mathbf{c}, \mathbf{d}\} \tag{9}$$

where $\mathbf{w}$, λ, σ, $\mathbf{c}$, and $\mathbf{d}$, are each a set of Np vectors.

5   The Classification Phase

In the classification phase, we assume that the pigment parameters are now known, so that the problem reduces to fit the algae to be classified to the set of fixed, exactly known pigments. The energy function to be minimized with respect to A only is given by:

$$E_c(A) = -\sum_{i=1}^{Na}\sum_{\lambda=1}^{\Lambda}\left(xa_i(\lambda) - \sum_{j=1}^{Np}A_{ij}s_j(\lambda)\right)^2 - log(P(A)) \tag{10}$$

where the $\mathbf{s}$ are considered fixed base signals.

To ameliorate the fitting, the base set can be incremented with a new, unknown "residual" pigment, still in the form of Gaussian mixture, to be estimated together with the coefficients of A. Since the "residual" pigment is likely to be different from alga to alga, in this case the classification system will be fed with a single alga, and the energy function becomes:

$$E_c(A, \mathbf{w}, \lambda, \sigma) = -\sum_{\lambda=1}^{\Lambda}\left(xa(\lambda) - \sum_{j=1}^{Np}A_j s_j(\lambda) - A_{Np+1}\sum_{k=1}^{NG}w_k exp\left(\frac{-(\lambda - \lambda_k)^2}{2\sigma_k^2}\right)\right)^2$$
$$- log(P(A)) - logP(\mathbf{w}, \lambda, \sigma) \tag{11}$$

The minimization of eq. 11 should be performed with respect to A and $\{\mathbf{w}, \lambda, \sigma\}$, where $\{\mathbf{w}, \lambda, \sigma\}$ indicate now the vectors of the weights, means and standard deviations, respectively, of the NG Gaussian components of the "residual" pigment.

It is worth noting that the "residual" pigment is not a "true" pigment, but an algae component that accounts for all the extra pigments contained in the alga in addition to the base pigments.

6   The Estimation Schedule

In both the learning and classification phases, the joint minimization of the energy functions of eqs. 8 and 11 (or the equivalent maximization - MAP estimation - of the related posterior probabilities) with respect to all the parameters is very computationally intensive. Thus, it is usually approached by means of alternating componentwise maximization with respect to groups of variables in turn. In this case, we subdivide the parameters Ω into three groups, assumed mutually independent, and, with reference to the learning phase, we adopt the following scheme:

$$\hat{A} = arg \max_{A} P(\mathbf{xa}|\mathbf{s}, A)P(A) \tag{12}$$

$$(\hat{\mathbf{w}}, \hat{\mu}, \hat{\sigma}) = arg \max_{\mathbf{w}, \mu, \sigma} P(\mathbf{xa}|\mathbf{s}, A)P(\mathbf{s})P(\mathbf{w}, \mu, \sigma). \tag{13}$$

$$(\hat{\mathbf{c}}, \hat{\mathbf{d}}) = arg \max_{\mathbf{c}, \mathbf{d}} P(\mathbf{s})P(\mathbf{c}, \mathbf{d}). \tag{14}$$

where the priors $P(A)$ and $P(\mathbf{s})$ are chosen in such a way to probabilistically enforce the over-mentioned constraints. We solve the above scheme by an overall Simulated Annealing scheme, where, at each steps, we alternate a Metropolis algorithm for separately estimating A and $\{\mathbf{c}, \mathbf{d}\}$, with a deterministic update for $\{\mathbf{w}, \lambda, \sigma\}$, based on gradient ascent.

7   Experimental Results

Absorbance maxima of the spectra of all algae divisions and plants are concentrated in the blue portion of the visible spectrum (and to lesser extent, in the red portion) where absorbance of chlorophyll a is higher. Absorbance attributable to accessory pigments is often difficult to quantify. For this reason, the ability to discriminate among distinct phylogenetic groups and potential species will depend upon the robustness of the technique chosen to differentiate the diverse components within the portion of the spectrum where carotenoids, phycobiliproteins and other chlorophylls absorb.

The learning phase of the classification method described in the paper was applied taking as input the spectra of 16 algae chosen to be representative of the 11 algal divisions, and the template spectra of 19 principal pigments among chlorophylls, carotenoids, phycobiliproteins and cytochromes. This phase resulted in the estimation of the Gaussian bands that compose each pigment, corrected for the distortions affecting the template pigments with respect to the "natural" ones, and the classification of the 16 algae. We assumed a maximum of 12 Gaussian bands for each pigment. At least in principle, the envelopes of the 19 pigments estimated in this phase can be considered reliable for successive classification of any other alga.

As per the classification part, some biological constraints were enforced, regarding the coefficients of A, which must be in $[0, 1]$, the non-simultaneous presence of chlorophyll b and c and the presence of more than one carotenoid in a same alga. Although the possible contribution from extra pigments has not been considered in this case, the algae fitting based on the only 19 available pigments was always satisfactory, with a root mean squared error no higher than 0.01.

The computational cost of the learning phase was high, due to the many parameters to be estimated via simulated annealing, and the many local minima

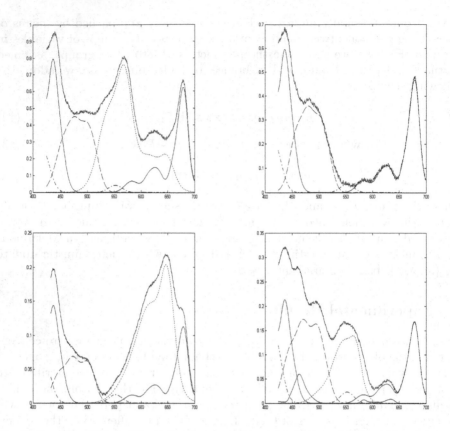

Fig. 1. Top left: fitting of Leptolyngbya sp.; Top right: fitting of Nannochloropsis sp.; Bottom left: fitting of Cyanophora paradoxa; Bottom right: fitting of Rhodomonas sp.. In all diagrams, solid line represents chlorophyll (a in the first three algae, a and c in the last one), dotted line represents the envelope of the phycobiliproteins, dashed line represents the envelope of the carotenoids, and, finally, dashdotted line represents cytochromes.

of the energy function which required the adoption of a very slow annealing scheme. However, as already said, the learning phase can be performed only once, while for the classification phase one alga at a time is feeded to the algorithm, so that the 19 coefficients of A alone need to be estimated.

From the analysis of the classification results of the 16 considered algae, we can state that chlorophyll a, chlorophyll b, chlorophyll c group and cytochromes, always have the same absorption maxima in all the algal division, though the relative intensity of their absorption bands can be different. On the other hand, carotenoids and phycobiliproteins show a variable pattern of peaks which is characteristic of each algal division. The results of the analysis of four of the different absorption spectra of the algal representatives are shown in Figure 1. The spectrum recorded on the trichomes of Leptolyngbya (Cyanophyta) (Figure 1, top left) shows that the main pigments of this cyanobacterium are chlorophyll

a, β-carotene (55% of the total amount of carotenoids), and among the phycobiliproteins, phycoeritrocyanin, c-phycocyanin and allophycocyanin are present.
In the case of Nannochloropsis, (Heterokontophyta) , only violaxantin is the main
carotenoid (about 50% of the total amount), and the phycobiliproteins are absent
(Figure 1, top right). In the spectrum of Cyanophora paradoxa (Glaucophyta)
(Figure 1 bottom left) only chlorophyll a is present, together with carotenoids,
and phycobiliproteins. Zeaxanthin is the main carotenoid (about 50% of the total amount), and the phycobiliproteins are represented by c-phycocyanin and
allophycocyanin. In Rhodomonas sp. (Cryptophyta) (Figure 1, bottom right),
both chlorophyll a and c are present, and the main carotenoid is alloxanthin
(about 25% of the total amount). Among the phycobiliproteins, phycoerithrin is
the only one detected, with the absorption maximum at 545 nm.

8   Conclusions

Absorbance attributable to accessory photo pigments often is difficult to quantify
and routinely discern for great number of these. For that reason, the ability to
discriminate among distinct phylogenetic groups (and potential species) will be
dependent upon the effectiveness of the techniques chosen to finding out "real"
Gaussian bands and their successive groupings. The microspectrophotometric
absorbance spectra and the successive Gaussian bands decomposition and their
fitting into the spectral envelope of pigment templates provides a realistic discrimination between the different algal divisions. For each division sensu Van
den Hoeck [1] the contribution and the spectral position of each pigment category is well defined. Even taking into account the inevitable approximations
we made in the template reconstructions, we consider our list of spectra the
most complete description of pigment distribution in the algal divisions grown
under laboratory condition. It can constitute a reference for different growing
conditions and diverse natural habitats.

References

1. Van Den Hoek, C., Mann, D.G., Jahns, H.M.: Algae - An Introduction to Phycol-
   ogy. Cambridge University Press, Cambridge (1995)
2. Keeling, P.J.: Diversity and evolutionary history of plastids and their hosts. Am.
   J. Bot. 91, 1481–1493 (2004)
3. Gualtieri, P.: Microspectroscopy of photoreceptor pigments in flagellated algae.
   Crit. Rev. Plant Sci. 9, 475–495 (1991)
4. French, C.S., Brown, J.S., Lawrence, M.C.: Four universal form of chlorophyll a.
   Plant Physiol. 49, 421–429 (1972)
5. Gulyayev, B.A., Litvin, F.F.: First and second derivatives of the absorption spectrum of chlorophyll and associated pigments in cells of higher plants and algae at
   $20C$. Biofizika 15, 701–712 (1970)
6. Butler, W.L., Hopkins, D.W.: Higher derivative analysis of complex absorption
   spectra. Photochem. Photobiol. 12, 439–450 (1970)

7. Hoepffner, N., Sathyendranath, S.: Effect of pigment composition on absorption properties of phytoplankton. Mar. Ecol. Prog. Ser. 73, 11–23 (1991)
8. Aguirre-Gomez, R., Weeks, A.R., Boxall, S.R.: The identification of phytoplankton pigments from absorption spectra. Int. J. Remote Sensing 22, 315–338 (2001)
9. Aguirre-Gomez, R., Boxall, S.R., Weeks, A.R.: Detecting photosynthetic algal pigments in natural populations using a high-spectral-resolution spectroradiometer. Int. J. Remote Sensing 22, 2867–2884 (2001)
10. Evangelista, V., Barsanti, L., Passatelli, V., Frassanito, A., Gualtieri, P.: Microspectroscopy of the Photosynthetic Compartment of Algae. Photochem. Photobiol. 82, 1039–1046 (2006)
11. Barsanti, L., Evangelista, V., Vesentini, C., Passarelli, V., Frassanito, A., Gualtieri, P.: Absorption microspectroscopy, theory and application in the case of photosynthetic compartment. Micron 38, 197–213 (2007)
12. Amari, S., Cichocki, A.: Adaptive blind signal processing -neural network approaches. Proceedings of the IEEE 86, 2026–2048 (1998)
13. Barros, A.K.: The independence assumption: dependent component analysis. In: Girolami, M. (ed.) Advances in ICA, ch. 4, Springer, London (2000)
14. Bedini, L., Herranz, D., Salerno, E., Baccigalupi, C., Kuruoglu, E.E., Tonazzini, A.: Separation of correlated astrophysical sources using multiple-lag data covariance matrices. EURASIP J. on Applied Signal Processing 15, 2400–2412 (2005)
15. Bell, A.J., Sejnowski, T.J.: An information maximization approach to blind separation and blind deconvolution. Neural Computation 7, 1129–1159 (1995)
16. Cichocki, A., Amari, S.: Adaptive Blind Signal and Image Processing. Wiley, New York (2002)
17. Hyvarinen, A.: Gaussian moments for noisy independent component analysis. IEEE Signal Processing Letters 6, 145–147 (1999)
18. Hyvarinen, A., Karhunen, J., Oja, E.: Independent Component Analysis. Wiley, New York (2001)
19. Salerno, E., Bedini, L., Kuruoglu, E., Tonazzini, A.: The problem of source separation in astrophysical images. In: Zharkova, V.V., Jain, L.C. (eds.) Artificial intelligence in recognition and classification of astrophysical and medical images, vol. SCI 46, pp. 200–209. Springer, Heidelberg (2007)
20. Tonazzini, A., Bedini, L., Salerno, E.: Independent component analysis for document restoration. Int. J. on Document Analysis and Recognition 7, 17–27 (2004)
21. Tonazzini, A., Bedini, L., Salerno, E.: A Markov model for blind image separation by a mean-field EM algorithm. IEEE Trans. on Image Processing 15, 473–482 (2006)

Semi-automatic Semantic Tagging of 3D Images from Pancreas Cells

Suzanne Little[1], Ovidio Salvetti[2], and Petra Perner[1]

[1] Institute of Computer Vision and Applied Computer Sciences, Germany
Suzanne.Little@ibai-institut.de, pperner@ibai-institut.de
[2] ISTI-CNR, Pisa, Italy
Ovidio.Salvetti@isti.cnr.it

Abstract. Detailed, consistent semantic annotation of large collections of multimedia data is difficult and time-consuming. In domains such as eScience, digital curation and industrial monitoring, fine-grained high-quality labeling of regions enables advanced semantic querying, analysis and aggregation and supports collaborative research. Manual annotation is inefficient and too subjective to be a viable solution. Automatic solutions are often highly domain or application specific, require large volumes of annotated training corpi and, if using a 'black box' approach, add little to the overall scientific knowledge. This article evaluates the use of simple artificial neural networks to semantically annotate micrographs and discusses the generic process chain necessary for semi-automatic semantic annotation of images.

Keywords: multimedia semantic annotation, semantic gap, artificial neural networks.

1 Introduction

Semantic annotation of media is recording high-level descriptive terms about the content of the media. It may be coarse-grained (descriptions at the image level) or fine-grained (descriptions at the segment or region level). Manual annotation (i.e., annotation by a human expert) is expensive, time-consuming, inconsistent and subjective. Tools and algorithms are available that can automatically extract low-level feature data from media objects such as color, shape, size, trajectory etc. However, these features are insufficient to support the queries required by domain experts who prefer to access data using higher-level terms such as catalyst, gaseous microemboli or mitochondria. Bridging the distance (often called the "multimedia semantic gap") between automatically extracted low-level features and high-level semantic terms is the focus of a great deal of research.

The concept of the "semantic gap" was initially defined in the field of psychology [1] and refers to the distance between information that can be extracted from the visual data and the interpretation that different users have for this same data. The difficulty is that humans cannot always describe what they see or explain why they interpret it in a certain way. Also there is rarely complete agreement between a group

P. Perner and O. Salvetti (Eds.): MDA 2006/2007, LNAI 4826, pp. 69–79, 2007.

of users about the interpretation of media content. Because of this, choosing the algorithms to extract low-level information from the media is also more complicated as we do not know what the key features are and therefore what procedure to apply.

Overcoming or mitigating this semantic gap to enable rapid and accurate semantic annotation of images is of particular importance in domains such as biology, geology, astronomy and industrial monitoring. Fields like these use high-resolution, high-throughput sensors and analytical machines to produce very large volumes of media content. Researchers need to be able to analyze and label objects of interest within these images and manage the quantity of data they provide.

This article discusses an approach to addressing the semantic gap using artificial neural networks in the context of the requirements for semi-automatic semantic annotation of scientific media. It summarizes of some of the main approaches to bridging the semantic gap (section 2) and outlines the motivations and characteristics of a sample applications (section 3). We present a generic processing chain for classifying images (section 4) and discuss the key challenges. Section 5 describes the use of artificial neural networks to classify image regions of a 3D slice sequence and presents the results of an initial evaluation.

2 Related Work

A variety of research efforts have investigated the use of techniques to extract semantic labels from different low-level visual or audio features. These approaches range from interfaces to facilitate user-driven annotation to systems integrating formalized knowledge structures, prototype based applications and automatic classification using machine-learning technologies. This section presents some examples of the more common approaches.

Interfaces that assist users to annotate or to link semantic terms to examples or sets of visual features are one approach to addressing the semantic gap. M-Ontomat-Annotizer [2] from the aceMedia project provides a graphical user interface for experts to link ontologies with low-level media features. The Rules-By-Example interface [3,4] also enables expert users to define mappings from low-level MPEG-7 features to high-level semantic terms using semantic web technologies such as OWL and RuleML/SWRL.

More traditional approaches have used machine-learning techniques such as statistical analysis, hidden markov models or artificial neural networks to determine semantic terms based on sets of low-level features. Chang et al. [5] applied a library of examples approach, which they call semantic visual templates. Zhao and Grosky [6] employ a latent semantic indexing technique which integrates a variety of automatically extracted visual features (global and sub-image color histograms and anglograms for shape-based and color-based representations) to enable semantic indexing and retrieval of images. Adams et al. [7] manually annotate atomic audio and video features in a set of training videos and from these develop explicit statistical models to automatically label the video with high-level semantic concepts. Work by Naphade et al. [8,9] proposed a statistical factor graph framework to bridge the gap between low-level features and semantic concepts. IBM alphaWorks have developed a tool, MARVEL [10,11], for "Multimedia Analysis and Retrieval" that applies heuristic techniques to automatically label image and video repositories based upon semantic models derived from sets of training examples.

Colantonio et al. [12] and Di Bona et al. [13] use neural networks to segment and characterize medical images. Previous work by Perner has shown that neural networks to tend generalize better than other methods [14] and can model non-linear decision surfaces. However, neural networks require a labeled training set of suitably significant size and variation. In contrast, a decision tree is a method that can easily be trained but they do not generalize as well as neural nets. If the variation in the data is very high then the preferred method would be case based reasoning [15,16]. This method does not necessarily generalize; it relies on samples and can incrementally learn. Unlike neural networks, both decision trees and case-based reasoning have the capability to explain their classifications.

The use of taxonomies or ontologies, either in combination with a user interface or with machine-learning approaches, enables reasoning using the relationships defined in the ontology and query expansion for better searching over the annotated dataset. Benitez and Chang [17] exploit structural semantics to label media objects by using WordNet [18] as a source of keywords in combination with Bayesian networks to provide media classification. Hollink et al. [19] use multiple ontologies to support the annotation of art images and applied a similar approach [20] to the annotation of news videos by combining links between visual features in a multimedia ontology (MPEG-7) and general semantic concepts defined in Wordnet. Bloehdorn et al. [21] have also used an MPEG-7 based ontology to formalise the relationships between high- and low-level visual features and semantic terms by recording 'prototype' instances that define the visual feature values.

The commonalities between these different approaches include the requirement for appropriate extraction of sets of low-level media features, the need for high-level, preferably well-defined, semantic terms to label and classify the media. The approaches that employ machine-learning techniques usually require a significant set of examples for training. The next section of this article presents an example application that needs support for automatic or semi-automatic annotation of image regions but is not able to initially supply a training set of statistically significant size.

3 Application

At the Institute for Molecular Bioscience of the University of Queensland, the Visible Cell project [22, 23] aims to increase understanding of the mammalian cell via the synthesis of physical data, models, mathematical and statistical simulations, and bioinformatics data. The objective of the project is to provide a visualization environment that seamlessly embeds macro-molecular structures, networks and quantitative simulations based on mathematical and complex-system models into a 3D mammalian cell reconstructed from high resolution tomograms and electron micrographs.

Figure 1 illustrates the data in this application which consists of 2D micrographic digital images of thin slices taken sequentially through a cell from a pancreas (a 3D object). We had 31 digital images from one cell. These 2D slices are used to create a 3D model of the sub-cellular structure. To build this 3D model and support the type of

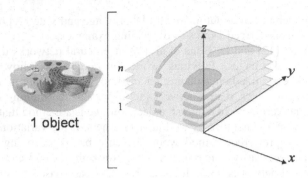

Fig. 1. 2D slices of a 3D object[1]

integrated, semantic queries desired by users, we need to label the single sub-cellular objects within the 2D slices. The objects of interest in this project are: mitochondria, ribosome, golgi stack, endoplasmic reticulum, mature granules, tubular vesicles.

A sample image is shown in Section 4.2, Figure 2. Finding an automatic image segmentation algorithms for images like this one is not easy since the cell is highly structured. Parallel research at the IMB [24] is working on algorithms for automated segmentation. However, until efficient and accurate automatic segmentation algorithms are able to be developed the method of choice is to manually label the objects. This means that a human user is sitting in front of a computer and circles objects of interest they can detect. The correct label for each object is not always obvious to the user and can only be determined after examining other images in the 3D stack to find adjacent regions that could be more easily identified. The user was able to label 7580 regions in the set of 31 slices. This is only the regions that the user was able to manually circle and identify – we cannot be certain that the user was able to identify all possible objects correctly.

To correctly label all these regions is time-consuming and difficult for a human user. Therefore an approach to automate part or all of the procedure is necessary. Initially, work using semantic inferencing rules was extended to evaluate the Rules-By-Example interface [3,26]. In this article we use artificial neural networks (ANN) since they are widely used in image classification [14] to discuss the generic process for semantic annotation and classification of images.

Our dataset consists of 5548 entries. Of the 7580 regions labeled by the expert, 2032 were identified using class labels that were not used in this evaluation being of minor objects that were of less interest. Each entry represents a region with the set of extracted low-level features and the user-assigned label. The distribution of the classes is shown in Table 1. The evaluation of the network, in terms of the accuracy rate, was done by test and train. The dataset was divided into sets of 70% for training and 30% for testing. The neural network architecture used was a simple feedforward network trained using a backpropagation algorithm. The features in the dataset are: Area, ConvexArea, FilledArea, MajorAxisLength, MinorAxisLength, Eccentricity, Solidity, Extent and DominantColor. Excluding DominantColor, which was

---

[1] 3D cell structure adapted from http://commons.wikimedia.org/wiki/Image:Biological_cell.svg

calculated separately, these terms are extracted using MATLAB's regionprops function from the Image Processing Toolbox. Through discussion with the expert users, general features such as size and shape were noted as being of particular importance and usefulness in distinguishing the different objects in the cell micrograph. This process of knowledge acquisition is not an easy procedure and requires experience to be able to extract relevant and useful knowledge from expert users. A methodology for doing so is presented in [27].

Table 1. Overall Sample Distribution (test and training sets)

	Endoplasmic Reticulum	Golgi	Mitochondria	Mature Granules	Ribosome	Tubular Vesicle
Number of Objects	2486 (45%)	1463 (26%)	221 (4%)	105 (2%)	1125 (20%)	148 (3%)

Future work includes converting the shape features used here to the more general MPEG-7 Shape Descriptors based on moments. MPEG-7 Region-base Shape Descriptors [31] would have been useful in this instance but extractors for generating them were not available at the time. The MPEG-7 standard itself does not bridge the gap between the low-level features and the higher semantic terms. Using standard MPEG-7 features helps with interoperability and provides an abstraction hierarchy or taxonomy of the different features for viewing or reasoning. The standard does not provide this hierarchy, however work on ontologies that incorporate the MPEG-7 feature terms include [28,29,30] and aims to achieve this level of structure. These proposed ontologies also add more intermediate terms to the media description vocabulary.

4 (Semi-)Automatic Annotation

The previous section presented an example application that would benefit from the ability to semi-automatically semantically annotate images. This section discusses the different levels of "semantic" labels and then describes the generic semantic annotation process for images.

4.1 Semantic Labels

Figure 2 shows the different types of semantic labels that can be applied to an image and gives some examples of the features and some possible values. The automatically or semi-automatically extracted *low-level* features have numerical values and, as such, are not easily understood or interpreted by a human user. The descriptors defined in the MPEG-7 standard [31] are examples of this type of feature. However, MPEG-7 does not give us a sufficient level of semantic terms for the visual features. It concentrates on descriptors such as region-based shape, scalable color or homo-geneous texture, etc.

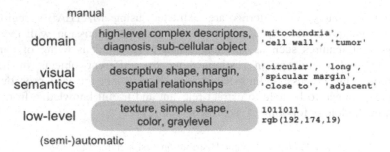

manual

domain	high-level complex descriptors, diagnosis, sub-cellular object	`'mitochondria',` `'cell wall', 'tumor'`
visual semantics	descriptive shape, margin, spatial relationships	`'circular', 'long',` `'spicular margin',` `'close to', 'adjacent'`
low-level	texture, simple shape, color, graylevel	`1011011` `rgb(192,174,19)`

(semi-)automatic

Fig. 2. Overview of different levels of semantic annotations

Of more use for human interpretation are *visually descriptive* or *semantic* features which describe characteristics in more usable, often symbolic vocabularies and may be drawn from standards such as BIRADs [32]. The highest-level of semantic labels are *domain specific* descriptors. These may be defined in a domain ontology, taxonomy or standard such as MeSH [33] or GO [34] and are rich, descriptive terms about the content of the image.

The visual semantic terms can be used to define mappings to the domain level terms. For example, "if object is long and thin and close to an object that is identified as 'Golgi' then the object is a 'Golgi' ".

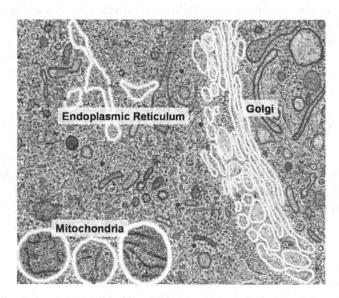

Fig. 3. Sample segmented pancreas cell micrograph, example regions for Endoplasmic Reticulum, Golgi and Mitochondria have been highlighted

In the example application described in section 3, the 2D slices contain objects that are described by human experts using both intermediate terms and labeled or classified using domain terms. For example, the sections labeled 'mitochondria' in Figure 3 have an uneven texture with distinct internal striations, they are circular in shape and generally large in area. By contrast the regions which make up the structure labeled the golgi stack have less distinctive visual features in common; they are generally smooth in texture, often long and thin in shape. Their most distinctive visual characteristic is their spatial relationships as the regions tend to lie in long, parallel alignments close to each other. A further example of the importance of spatial relationships is the endoplasmic reticulum regions which are only distinguishable by the presence of the small ribosome region touching it. As the ribosome generally only touches the endoplasmic reticulum object at one point in 3D space, it is often only able to be identified in a small subset of the 2D slices. Its presence in other slices is inferred through the adjacency of regions along the z-plane.

As you can see these descriptions are not numerical values but rather intermediate descriptors used to describe the objects of interest in terms of their shape, texture and location within the complete scene. In order to generate the high-level domain terms, useful for querying and to assist in generating the final 3D models, semantic mappings need to be developed from the low-level to the intermediate terms and then from the intermediate to the domain level descriptions. The next section describes a generic process chain for image annotation and classification.

4.2 The Generic Process Chain Necessary for Semi-automatic Semantic Annotation

The process of semantically annotating images based on low-level features usually follows a common abstract procedure. The generic processing chain for image understanding is shown in Figure 4.

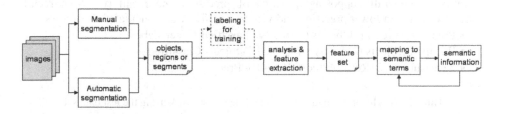

Fig. 4. Generic process chain for image annotation or classification

The first step is segmentation where the image is segmented into background and objects of interest. Ideally algorithms for automatic segmentation should be used to analyze and divide the images. However, devising a general procedure for segmenting images is not always possible and sometimes manual or semi-automatic processes are required.

Once we have determined what image pixels belong to which object, we need to label the regions representing the objects. Using these regions, we can then extract the object features (e.g., low-level features such as graylevel, simple shape features, texture and color) to produce a feature set for each object.

These low-level features need to be mapped to intermediate semantic terms such as 'circular', 'long', 'angular margin', 'spicular margin', etc. These semantic terms, more familiar to an expert user, can be used to describe the class or category of an object in the image. This is the first phase of mapping to semantic terms.

To achieve fine-grained, high-level descriptions of objects is sometimes only possible by taking into account intermediate descriptors such as spatial information about location or relation to other objects within the image. Coarse-grained descriptions of the complete scene requires the grouping of objects and describing their spatial relation to each other. The intermediate level semantic terms need to be mapped to domain level terms which describe the content depicted in the image. This is the second phase of mapping to semantic terms.

There is no universal algorithm available that can automatically process the semantic information for all kinds of images. This means that specific images need special processing functions in order to implement the processing chain described in this section.

5   Results and Discussion

Table 2 shows the evaluation results of the neural network using the test data set of 1627 objects. *User Labeled* is the number of objects in the test set that were labeled as a specific 'class' by the expert user. *Objects Classified* is the number of objects in the test set that the network classified as 'class' while *Objects Correctly Classified* is the number of the Objects Classified whose class label corresponds with that given by the expert user. *Precision*, *Recall* and *f-measure* are terms from the information retrieval domain [25]. Precision and recall are both measurements of classification quality. f-measure provides a more useful measure of the overall performance since it takes into account the generally opposing qualities of precision and recall (i.e. high recall generally results in lower precision and vice versa). They are calculated as follows.

- Precision (Prec.) is Objects Correctly Classified / User Labeled
- Recall is Objects Correctly Classified / Objects Classified
- f-measure is (2*Recall*Prec.)/(Recall + Prec.)

Table 2. Results of simulating the trained neural network using the test data set

Object Class	User Labeled in Test dataset	Output of Neural Network	Objects Correctly Classified	Prec.	Recall	f-measure
Endoplasmic Reticulum	435	482	226	0.469	0.520	0.493
Golgi	696	928	474	0.511	0.681	0.584
Mitochondria	38	30	30	1.000	0.789	0.882
Mature Granules	63	67	47	0.701	0.746	0.723
Ribosome	354	305	291	0.954	0.822	0.883
Tubular Vesicles	41	28	26	0.929	0.634	0.754

The accuracy of the system is calculated as the total number of correctly classified objects (1094) divided by the total number of objects (1627). This gives an accuracy rate of 0.627.

The high precision for Mitochondria and to a lesser extent for Ribosome and Tubular Vesicles indicates the better visual distinction based around the shape of these objects – Mitochondria are large and tend to be more circular (higher Eccentricity values); Ribosome are small and more irregular in shape while Tubular Vesicles are circular (very high Eccentricity values) and consistently very small in size.

The poor performance in identifying Endoplasmic Reticulum and Golgi objects is possibly due to their low visual distinction when only considering basic shape features. As section 4.1 discussed, they are much more easily described using texture and more intermediate descriptors such as spatial relations.

Neural networks are generally better at discriminating between classes, as is shown in [14]. Therefore the difficulty this network has in distinguishing between Golgi and Endoplasmic Reticulum is interesting. However, we feel this is attributable to the lack of spatial relations in the input features.

We didn't achieve the accuracy that we were hoping for since we mapped directly from the low-level features to the class label. In addition we did not have a large enough set of features such as spatial relationship to other objects. This meant that information about spatial relations and intermediate shape and texture descriptors were not incorporated into the classification process.

Also the testing and training data all came from a single example cell. This means that this dataset might not represent adequate statistical variation among the data from different cells. It is not clear that this network would perform as well using data from another cell. When new data is added, the ANN needs to be retrained in order to achieve good performance. Neural nets do not support incremental learning. Other methods such as decision trees or case-based reasoning are preferred for this reason.

Overall, the small sample set and the limited variation in the source object (one cell) restricts the statistical significance of these results and the conclusions that can be made from this evaluation. However, the result from this network provide support for the view that mapping from low-level features to intermediate terms and then from intermediate terms to domain descriptions is likely to be a more successful approach.

6 Conclusion and Future Work

Using artificial neural networks to map from low-level media features directly to high-level semantic terms for image regions does not demonstrate a particularly high level of accuracy. While previous applications have shown that neural networks can be effective in image classification tasks [12,13,14], we believe that a multi-stage process, as proposed in section 4.2, is likely to be more effective for semantic annotation of image regions.

However, until efficient and accurate automatic segmentation algorithms are able to be developed, techniques are needed for semi-automatic semantic annotation that can handle small input data sets, evolving models and rapidly increasing data. Therefore, we aim to develop a system that can operate on an initial, small dataset but incrementally adapt and improve with the addition of further data as it becomes available. The classification system will eventually become more generalized and have improved accuracy as new cell slices are incorporated into the data set. This situation is common in many medical and scientific research fields where the available experimental data may initially be relatively small but which will increase as further experiments and analysis are conducted.

We believe that to generate a semantic description of an image you can not use the low-level features directly, you have to first map them to intermediate symbolic or semantic terms that make sense for a domain expert. Therefore we intend to focus on a multi-step classification procedure where intermediate terms (such as 'circular', 'fine speckled margin', 'adjacent' etc.) are created from the automatically extracted low-level features. These terms can then be used to build better classification systems using techniques such as inferencing rules [2,3], case-based reasoning [15,16] or decision trees [27].

Acknowledgments. This project has been sponsored by the European Commission within the project "Multimedia Understanding through Semantics, Computation, and Learning, MUSCLE" No. 507752.

References

1. Finke, R.: Principles of Mental Imagery, pp. 89–90. MIT Press, Cambridge (1989)
2. Saathoff, C., Petridis, K., Anastasopoulos, D., Timmermann, N., Kompatsiaris, I., Staab, S.: M-OntoMat-Annotizer: Linking Ontologies with Multimedia Low-Level Features for Automatic Image Annotation. In: Sure, Y., Domingue, J. (eds.) ESWC 2006. LNCS, vol. 4011, Springer, Heidelberg (2006)
3. Little, S., Hunter, J.: Rules-By-Example - a Novel Approach to Semantic Indexing and Querying of Images. In: McIlraith, S.A., Plexousakis, D., van Harmelen, F. (eds.) ISWC 2004. LNCS, vol. 3298, Springer, Heidelberg (2004)
4. Hunter, J., Little, S.: A Framework to enable the Semantic Inferencing and Querying of Multimedia Content. International Journal of Web Engineering and Technology (IJWET) Special Issue on the Semantic Web 2 (December 2005)
5. Chang, S.F., Chen, W., Sundaram, H.: Semantic Visual Templates: linking visual features to semantics. In: ICIP 1998, Chicago, Illinois (1998)
6. Zhao, R., Grosky, W.: Negotiating The Semantic Gap: From Feature Maps to Semantic Landscapes. Pattern Recognition 35(3), 51–58 (2002)
7. Adams, B., Iyengar, G., Lin, C., Naphade, M., Neti, C., Nock, H., Smith, J.: Semantic Indexing of Multimedia Content Using Visual, Audio and Text Cues. EURASIP Journal on Applied Signal Processing (2003)
8. Naphade, M., Kozintsev, I., Huang, T., Ramchandran, K.: A Factor Graph Framework for Semantic Indexing and Retrieval in Video. In: CBAIVL 2000, IEEE Computer Society Press, Los Alamitos (2000)
9. Naphade, M., Huang, T.: Detecting semantic concepts using context and audiovisual features. In: Proceedings of Detection and Recognition of Events in Video, 2001, pp. 92–98 (2001)
10. IBM alphaWorks. Multimedia Analysis and Retrieval Engine (MARVEL). Last accessed (August 2006), http://www.alphaworks.ibm.com/tech/marvel
11. Natsev, A., Naphade, M., Tesic, J.: Learning the Semantics of Multimedia Queries and Concepts from a Small Number of Examples. ACM Multimedia (2005)
12. Colantonio, S., Gurevich, I.B., Salvetti, O.: Automatic Fuzzy-Neural based Segmentation of Microscopic Cell Images. In: Industrial Conference on Data Mining - Workshops, pp. 34–45 (2006)

13. Di Bona, S., Niemann, H., Pieri, G., Salvetti, O.: Brain volumes characterisation using hierarchical neural networks. Artificial Intelligence in Medicine 28(3), 307–322 (2003)
14. Perner, P., Zscherpel, U., Jacobsen, C.: A comparison between neural networks and decision trees based on data from industrial radiographic testing. Pattern Recognition Letters 22, 47–54 (2001)
15. Perner, P. (ed.): Case-Based Reasoning on Images and Signals. Springer, Heidelberg (in print, 2007)
16. Perner, P.: Prototype-based classification. Journal of Applied Intelligence (in print, 2007)
17. Benitez, A., Chang, S.-F.: Image classification using multimedia knowledge networks. In: ICIP 2003, vol. 2, 3, pp. III-613–616 (2003)
18. Fellbaum, C.: Wordnet, An Electronic Lexical Database. MIT press, Cambridge (1998)
19. Hollink, L., Schreiber, A., Wielemaker, J., Wielinga, B.: Semantic Annotation of Image Collections. In: KCAP 2003, Florida, USA (2003)
20. Hollink, L., Worring, M., Schreiber, A.: Building a Visual Ontology for Video Retrieval. In: ACM Multimedia, Singapore (November 2005)
21. Bloehdorn, S., Petridis, K., Saathoff, C., Simou, N., Tzouvaras, V., Avrithis, Y., Handschuh, S., Kompatsiaris, I., Staab, S., Strintzis, M.G.: Semantic Annotation of Images and Videos for Multimedia Analysis. In: Gómez-Pérez, A., Euzenat, J. (eds.) ESWC 2005. LNCS, vol. 3532, Springer, Heidelberg (2005)
22. Hunter, J., Regan, M., Little, S.: Position Paper for Semantic Web Life Sciences Workshop – The Visible Cell. In: W3C's Semantic Web Life Sciences Workshop, Cambridge Mass. (2004)
23. Marsh, B., Mastronarde, D., Buttle, K., Howell, K., McIntosh, J.: Organellar relationships in the Golgi region of the pancreatic beta cell line, HIT-T15, visualized by high resolution electron tomography. Proceedings of the National Academy of Sciences of the United States of America 98(5), 2399–2406 (2001)
24. Institute for Molecular Biology, "Visible Cell Project", University of Queensland, Australia, http://www.visiblecell.com
25. van Rijsbergen, C.J.: Information Retrieval, 2nd edn. Butterworths (1979)
26. Hollink, L., Little, S., Hunter, J.: Evaluating the Application of Semantic Inferencing Rules to Image Annotation. In: KCAP 2005, Banff, Canada (2005)
27. Perner, P.: Data Mining on Multimedia Data. LNCS. Springer, Heidelberg (2002)
28. Hunter, J.: Adding Multimedia to the Semantic Web - Building and Applying MPEG-7 Ontology. In: Stamou, G., Kollias, S. (eds.) Multimedia Content and the Semantic Web: Standards, and Tools, Wiley, Chichester (2005)
29. Tsinaraki, C., Polydoros, P., Kazasis, F., Christodoulakis, S.: Ontology-based Semantic Indexing for MPEG-7 and TV-Anytime Audiovisual Content. Special issue of Multimedia Tools and Application Journal on Video Segmentation for Semantic Annotation and Transcoding 26, 299–325 (2005)
30. Garcia, R., Celma, O.: Semantic integration and retrieval of multimedia metadata. In: SemAnnot 2005, Galway, Ireland (November 2005)
31. Manjunath, B.S., Salembier, P., Sikora, T. (eds.): Introduction to MPEG-7: Multimedia Content Description Interface. Wiley, Chichester (2003)
32. American College of Radiology, Breast Imaging Reporting and Data System (BI-RADS®)
33. National Library of Medicine. "Medical Subject Headings (MeSH)." Last accessed (July 2007), http://www.nlm.nih.gov/mesh/
34. Smith, B., Williams, J., Schulze-Kremer, S.: The Ontology of the Gene Ontology. In: Proceedings of AMIA Symposium (2003)

Combinatorial Synthesis of Thin Mixed Oxide-Films and Examinations of Their Piezoelectricity by Ultrasonic Piezo-Mode Imaging

Daniela Rende[1], Wilhelm F. Maier[1], Kerstin Schwarz[2], Ute Rabe[2],
and Walter Arnold[2]

[1] Lehrstuhl für Technische Chemie, Saarland University, Bldg. C 4.2,
D-66123 Saarbrücken, Germany
[2] Fraunhofer Institute for Non-Destructive Testing (IZFP), Bldg. E 3.1,
University, D-66123 Saarbrücken, Germany

Abstract. The development of an automated production of thin films and the characterization of their piezoelectric properties in high-throughput is described. A library of 50 undoped as well as doped lead zirconate titanate $Pb(Zr,Ti)O_3$ (PZT) coatings was produced by sol deposition. Afterwards, the piezoelectric properties of the library films were analyzed by automated atomic force microscopy employing the ultrasonic piezo-mode.

Keywords: PZT, combinatorial chemistry, high throughput, sol-gel synthesis, AFM.

1 Introduction

There is in increasing interest in thin films of ceramics with defined mechanical, magnetic or electro-active properties with applications ranging from protective coatings, magneto-optical image storage devices, and micro-electro-mechanical systems (MEMS). A new approach using high-throughput technology to discover materials having defined ferroelectric properties is presented. For the investigation of new piezoelectric materials a lead zirconate titanate $Pb(Zr,Ti)O_3$ (PZT) sol-gel-synthesis was applied, which can be modified in a combinatorial way. The use of an acidic sol-gel-recipe allows a broad variation in composition and doping with many elements.

2 Synthesis

Single-element precursor solutions were mixed by a pipetting robot to form 17 undoped PZT sols of different composition by varying the ratio of lead to zirconium to titanium as piezoelectric reference materials and 33 PZT sols (Pb1.1 Zr0.58 Ti0.42) doped with 6 mol-% of different elements. Subsequently, the precursor solutions were deposited onto a pre-structured silicon wafer (**Figs. 1 & 2**) by chemical solution deposition (CSD) /1/.

P. Perner and O. Salvetti (Eds.): MDA 2006/2007, LNAI 4826, pp. 80–83, 2007.
© Springer-Verlag Berlin Heidelberg 2007

Fig. 1. Schematic view of the spread of the liquid on a structured substrate /2/

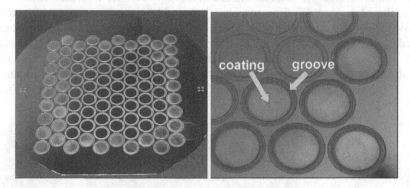

Fig. 2. (a) Micrograph of chemical solution deposition of a thin-film library on a structured 6"-silicon wafer, and (b) an enlarged image (3 × 3.5 cm²) of this library

3 Measurement Technique

The piezoelectric properties of the films on the library were automatically investigated by atomic force microscopy (AFM) employing the ultrasonic piezo-mode /3/. A sinusoidal voltage is applied between an electrically conductive cantilever and a base electrode below a piezoelectric sample (**Fig. 3**) /4/. The localized electric field emanating from the sensor tip of the AFM induces a deformation on a local scale due to the inverse piezoelectric effect. The ac frequency corresponds to the contact resonance of the conductive cantilever. The amplitude and phase of the induced surface displacement is sensed by the tip. The library was analyzed by the AFM tool "programmed move", which allows the cantilever to be positioned automatically to every of the samples on the library. The ultrasonic piezo-mode is an acoustic near-field technique which has been extensively discussed for example in a special session of the International Symposium on Acoustical Imaging (AI27) which was held in Saarbrücken in 2003 /5/.

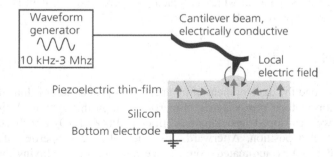

Fig. 3. Principle of the AFM ultrasonic piezo-mode

4   Results of Automated Analysis Using Ultrasonic Piezo-Mode

A topography image of the surface was acquired simultaneously with the acoustic image. The ultrasonic piezo-mode images of 17 piezoelectric reference films as well as 15 of the 33 doped PZT- films revealed dark and bright areas corresponding to high and low local vibration amplitudes. The samples doped with the following elements showed a contrast in the ultrasonic piezo-mode image: Ba, Ce, Co, Cu, Li, Sr, Na, K, La, Pr, Rb, Ag, Te, V(III), and V(V) indicative of piezoelectric properties. **Fig. 4** illustrates examples of topography images of three different samples ((a), (c), and (e)) together with their corresponding ultrasonic piezo-mode images ((b), (d), and (f)) /1/.

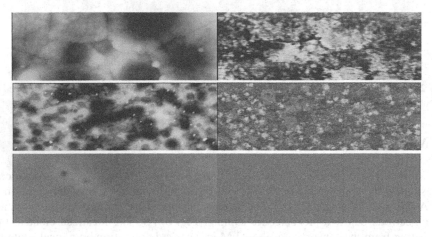

Fig. 4. (a), (c), and (e): Topography images of three different samples doped with Ba, Sr and Er, respectively. The image size is 10×3 µm². The scale covers 50 nm difference in height. (b), (d), (f): Corresponding ultrasonic piezo-mode images of the same surface areas as shown in (a), (c), and (e). The amplitude range of the ultrasonic piezo-mode images is 1.5 V.

The contrast in the ultrasonic piezo-mode image is caused by ferroelectric domains, but false positives resulting from variation of film roughness cannot yet be excluded. To exclude such topography effects, the samples have been examined by impedance spectroscopy as well yielding ε. For the PZT coating doped with Sr an ε-value of 194.36 was obtained which is typical for perovskites. These results confirm the ultrasonic AFM piezo-mode measurements.

5   Conclusion

We have described the development of an automated production of thin films and the characterization of their piezoelectric properties in high-throughput. A library of 50 undoped as well as doped lead zirconate titanate $Pb(Zr,Ti)O_3$ (PZT) coatings was produced by sol deposition. Afterwards, the piezoelectric properties of the library films were analyzed by automated atomic force microscopy employing the ultrasonic piezo-mode. The results show that the piezoelectric properties can be determined by using the atomic force microscopy employing the ultrasonic piezo-mode. For the automatic determination of the piezoelectric properties an image processing procedure is needed that will be developed in future.

Acknowledgements

This work was supported by grants from the German Science Foundation through the "SFB 277" as well as the "Graduate College 232 "Materials for efficient energy use". The authors thank B. Wessler from Siemens AG for the supply of structured wafers.

References

1. Rende, D., Schwarz, K., Rabe, U., Maier, W.F., Arnold, W.: Combinatorial Synthesis of Thin Mixed Oxide Films and Automated Study of their Piezoelectric Properties. Progr. Solid State Chem. 35, 361 (2007)
2. Wessler, B., Jéhanno, V., Rossner, W., Maier, W.F.: Combinatorial Synthesis of Thin Film Libraries for Microwave Dielectrics. Appl. Surf. Sci. 223, 30 (2004)
3. Rabe, U., Kopycinska, M., Hirsekorn, S., Arnold, W.: Evaluation of the contact resonance frequencies in atomic force microscopy as a method for surface characterization. Ultrason. 40, 49 (2002)
4. Güthner, P., Dransfeld, K.: Local Poling of Ferroelectric Polymers by Scanning Force Microscopy. Appl. Phys. Lett. 61, 1137 (1992)
5. Arnold, W., Hirsekorn, S. (eds.): Proc. 27th Int. Symp. Acoustical Imaging. Kluwer Academic Publishers, New York, Boston, Dordrecht (2004)

A Novel Image Feature for Nuclear-Phase Classification in High Content Screening

Tuan D. Pham[1] and Xiaobo Zhou[2]

[1] Bioinformatics Applications Research Center,
School of Mathematics, Physics, and Information Technology,
James Cook University,
Townsville, QLD 4811, Australia
[2] HCNR Center for Bioinformatics,
Harvard Medical School,
Boston, MA 02115, USA

Abstract. Cellular imaging is an exciting area of research in computational life sciences, which provides an essential tool for the study of diseases at the cellular level. In particular, to faciliate the usefulness of cellular imaging for high content screening, image analysis and classification need to be automated. In fact the task of image classification is an important component for any computerized imaging system which aims to automate the screening of high-content, high-throughput fluorescent images of mitotic cells. It can help biomedical and biological researchers to speed up the analysis of mitotic data at dynamic ranges for various applications including the study of the complexity of cell processes, and the screening of novel anti-mitotic drugs as potential cancer therapeutic agents. We propose in this paper a novel image feature based on a spatial linear predictive model. This type of image feature can be effectively used for vector-quantization based classification of nuclear phases. We used a dataset of HeLa cells line to evaluate and compare the proposed method on the classification of nuclear phases. Experimental results obtained from the new feature are found to be superior to some recently published results using the same dataset.

Keywords: Feature extraction, microscopic imaging, cellular classification, high content screening.

1 Introduction

By the use of fluorescence-based reagents, high content screening (HCS) studies cell functions by extracting the temporal and spatial information about target activities within cells [1]. Particularly due to the huge volumes of acquired images, the automation of HCS systems has become necessary to help life-science researchers understand the complex process of cell division or mitosis at a rapid speed [2,3]. Its power comes from the sensitivity and resolution of automated light microscopy with multi-well plates, combined with the availability of fluorescent probes that are attached to specific subcellular components, such as

P. Perner and O. Salvetti (Eds.): MDA 2006/2007, LNAI 4826, pp. 84–93, 2007.

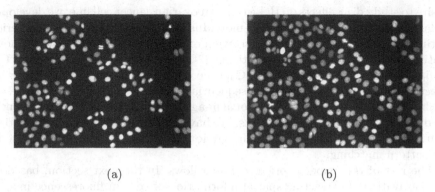

Fig. 1. Progress of cell division after a period of time – (a) typical fluorescence microscopy image of cells; (b) division of cells showing changes in temporal and spatial information

chromosomes and microtubules, for visualization of cell division or mitosis using standard epi-fluorescence microscopy techniques [17]. By employing a carefully selected reporter probes and filters, fluorescence microscopy allows specific imaging of phenotypes of essentially any cell component [13]. With these probes we can determine both the amount of a cell component, and most critically, its distribution within the cell relative to other components. Typically, 3-4 different components are localized in the same cell using probes that excite at different wavelengths. Any change in cell physiology would cause a redistribution of one or more cellular components, and this redistribution provides a certain cytological marker that allows for scoring of the physiological change.

In time-lapse microcopy images are usually captured in a time interval of more than 10 minutes. During this period dividing nuclei may move far away from each other and daughter cell nuclei may not overlap with their parents. Figure 1(a) shows a microscopic image of cells; whereas Figure 1(b) shows a typical example of the progress of cell division after some period of time.

An essential task for high content screening is to measure cell cycle progression (inter phase, prophase, metaphase, and telophase) in individual cells as a function of time. Cell cycle progress can be identified by measuring nuclear changes. Thus, automated time-lapse fluorescence microscopy imaging provides an important method for the observation and study of cellular nuclei in a dynamic fashion [6,10]. Stages of an automated cellular imaging analysis consist of segmentation, feature extraction, classification, and tracking of individual cells in a dynamic cellular population; and the classfication of cell phases is considered the most difficult task of such analysis [4].

Given the advanced fluorescent imaging technology, there still remain technical challenges in processing and analyzing large volumes of images generated by time-lapse microscopy. The increasing quantity and complexity of image data from dynamic microscopy renders manual analysis unreasonably time-consuming. Therefore, automatic techniques for analyzing cell-cycle progress are of considerable interest in the drug discovery process. Being motivated by the desire to study drug effects on HeLa cells, an ovarian cancer cell line, we developed a classification model for identifying individual cell phase changes during a period of time. We seek to extract the information of image spatial-continuity as a novel feature using the concept of geostatistics [7] and the theory of linear predictive coding (LPC) [12]. The theory of LPC has been successfully applied for extracting spectral feature of one-dimensional sequential data [16]. In this study we present a scheme for extracting two-dimensional image feature known as the spatial linear predictive coding (SLPC) coefficients. We have implemented the vector quantization (VQ) method [5] to design a template for each SLPC-based cell-phase model for pattern matching.

The rest of this paper is organized as follows. In the next section, based on the motivation of extracting spatial information of cells in fluorescence images, we will introduce the derivation of spatial LPC coefficients using the theory of geostatistics as a new type of feature of cellular images. We will then address how we have implemented the method of vector quantization to design the prototypes for cell phases based on the spatial LPC coefficients. Following the technical presentations, we will illustrate the performance of the proposed approach by testing and comparing the SLPC-VQ based classifier with real image data and other methods respectively. We will conclude the finding and suggest some issues for future research.

2   SLPC-Based Image Feature

Spatial linear predictive coding (SLPC) coefficients can be determined using the method of ordinary kriging (OK) introduced in the theory of geostatistics [9,7]. This geostatistical method tries to predict the unknown value using the following weighted linear combination of the available samples:

$$\hat{x}_0 = \sum_{i=1}^{k} w_i\, x_i \tag{1}$$

where $\hat{x}_0$ is the estimate of x_0, w_i is the weighting parameter that expresses the relationship between value x_0 and available value x_i.

There are different approaches for determining the weights to the available or neighbor data with respect to the unknown value, and different approaches lead to different computational schemes. One particular approach for computing these weights optimally is to minimize the average error of estimation. Let r_j denote the error between any particular estimated $\hat{x}_j$ value and the true value x_j:

$$r_j = \hat{x}_j - x_j \tag{2}$$

then the average error, denoted as r_a, of k estimates is

$$r_a = \frac{1}{k} \sum_{j=1}^{k} r_j \tag{3}$$

However, minimizing r_a is unrealistic because the true values $x_1, \ldots, x_k$ are not known. In other words, it is not possible to minimize the variance of the actual errors, but it is possible to minimize the variance of the modeled error which is defined as the difference between the random variables modeling the estimate and the true value. Such optimal solution for the weights w_i for all i can be obtained by kriging known as the best linear unbiased estimator. Let $V(x_1), \ldots, V(x_k)$ be the random variables for k samples $x_1, \ldots, x_k$ respectively; and $V(x_0)$ be the random variable for x_0. These random variables are assumed to have the same probability distribution, and the expected value of the random variables at all locations is $E\{V\}$. Thus, the estimate of x_0 is also a random variable and expressed by a weighted linear combination of the random variables at k locations:

$$\hat{V}(x_0) = \sum_{i=1}^{k} w_i \, V(x_i) \tag{4}$$

And the error of estimation is

$$R(x_0) = \hat{V}(x_0) - V(x_0) \tag{5}$$

Alternatively we have

$$R(x_0) = \sum_{i=1}^{k} w_i V(x_i) - V(x_0) \tag{6}$$

The expected value of the error of estimate is

$$E\{R(x_0)\} = \sum_{i=1}^{k} w_i E\{V(x_i)\} - E\{V(x_0)\} \tag{7}$$

Based on the assumptions of stationary random function and unbiased estimation, the variance of the error can be expressed as [7]

$$\sigma_R^2 = \sigma^2 + \sum_{i=1}^{k} \sum_{j=1}^{k} w_i w_j C_{ij} - 2 \sum_{i=1}^{k} w_i C_i \tag{8}$$

where C_{ij} stands for the covariance of x_i and x_j; and the above equation defines the variance of error as a function of $w_1, \ldots, w_k$.

An optimal choice for the weighting parameters $w_1, \ldots, w_k$ is to minimize σ_R^2 subject to $\sum_i w_i = 1$. A solution can be obtained by the Lagrange multiplier method:

$$\sigma_R^2 = \sigma^2 + \sum_{i=1}^{k} \sum_{j=1}^{k} w_i w_j C_{ij} - 2 \sum_{i=1}^{k} w_i C_i + 2\beta(\sum_{i=1}^{k} w_i - 1) \tag{9}$$

where β is a Lagrange multiplier.

The error variance term, σ_R^2, can now be minimized by differentiating (9) with respect to the weights and the Lagrange parameter, and setting each one to zero. By doing so, we obtain the following equations.

$$\sum_{j=1}^{k} w_j C_{ij} + \beta = C_{i0}, \ \forall i = 1, \ldots, k \tag{10}$$

$$\sum_{i=1}^{k} w_i = 1 \tag{11}$$

The above system of equations are known as the ordinary kriging system, which can be expressed in matrix notation as

$$\mathbf{C}\,\mathbf{w} = \mathbf{D} \tag{12}$$

where

$$\mathbf{C} = \begin{bmatrix} C_{11} & \cdots & C_{1k} & 1 \\ . & \cdots & . & . \\ . & \cdots & . & . \\ . & \cdots & . & . \\ C_{k1} & \cdots & C_{kk} & 1 \\ 1 & \cdots & 1 & 0 \end{bmatrix}$$

$$\mathbf{w} = \begin{bmatrix} w_1 & \cdots & w_k & \beta \end{bmatrix}^T$$

$$\mathbf{D} = \begin{bmatrix} C_{10} & \cdots & C_{k0} & 1 \end{bmatrix}^T$$

Thus the values of the weights can be obtained by solving

$$\mathbf{w} = \mathbf{C}^{-1}\,\mathbf{D} \tag{13}$$

The sample covariance used for the kriging estimator can be calculated as

$$C(h) = \frac{1}{N(h)} \sum_{(i,j)|h_{ij}=h} x_i x_j - \left(\frac{1}{n}\sum_{k=1}^{n} x_k\right)^2 \tag{14}$$

in which the sample covariance is a function of the lag distance h, $N(h)$ is the number of pairs that x_i and x_j are separated by h, and n is the total number of data. In this sense, it very convenient to include the spatial correlation of an image in different geometrical directions. Due to the structured format of pixels, one easy way is to compute $C(h)$ in the vertical and horizontal directions of the image.

Furthermore, on the derivation of the error of variance, it is assumed that the random variables have the same mean and variance which lead to the development of the mathematical relationship between the variogram, denoted as $\gamma(h)$, and the covariance [7]

$$\gamma(h) = \sigma^2 - C(h) \tag{15}$$

where the sample $\gamma(h)$ is defined as

$$\gamma(h) = \frac{1}{2N(h)} \sum_{(i,j)|h_{ij}=h} (x_i - x_j)^2. \tag{16}$$

3 Spatial LPC-VQ-Based Classification

Generally speaking, to calculate a distortion measure between two vectors $\mathbf{x}$ and $\mathbf{y}$, denoted as $D(\mathbf{x}, \mathbf{y})$, is to calculate a cost of reproducing any input vector $\mathbf{x}$ as a reproduction of vector $\mathbf{y}$. Given such a distortion measure, the mismatch between two signals can be quantified by an average distortion between the input and the final reproduction. Intuitively, a match of the two patterns is good if the average distortion is small. One popular distortion is the Itakura-Saito distortion measure [8]. It was pointed out that the Itakura-Saito distortion measure is connected with many statistical and information theories [16] including the likelihood ratio test, discrimination information, and Kullback-Leibler divergence. Based on the notion of the Itakura-Saito distortion measure, the spatial LPC likelihood ratio distortion between two signals s and s' is derived and expressed as

$$D_{LR} = \frac{\mathbf{w}'^T \mathbf{C}_s \mathbf{w}'}{\mathbf{w}^T \mathbf{C}_s \mathbf{w}} - 1 \tag{17}$$

where $\mathbf{C}_s$, which is expressed in (13), is the spatial covariance matrix of sequence s associated with its SLPC coefficient vector $\mathbf{w}$, and $\mathbf{w}'$ is the SLPC coefficient vector of signal s'.

Assume that we have a set of T feature vectors obtained the trained images, which are represented by the corresponding set of T SLPC vectors $\mathbf{W} = \{\mathbf{w}_1, \mathbf{w}_2, \ldots, \mathbf{w}_T\}$, where $\mathbf{w}_t = (w_{t1}, w_{t2}, \ldots, w_{tp})$. It can be seen that these SLPC vectors represent a type of feature of the image. To handle the problem of large data of feature vectors which is particularly common in high content screening, we seek to compress the feature space using a vectorized-signal compression technique of vector quantization. In fact vector quantization is one of the most suitable methods for compressing LPC-based vectors [12,16].

Let the codebook of the SLPC vectors be $\{\mathbf{c}_1, \mathbf{c}_2, \ldots, \mathbf{c}_N\}$, where $\mathbf{c}_n = (c_{n1}, c_{n2}, \ldots, c_{np})$, $n = 1, 2, \ldots, N$ are codewords. Each codeword $\mathbf{c}_n$ is assigned to an encoding region R_n in the partition $\Omega = \{R_1, R_2, \ldots, R_N\}$. The source LPC vector $\mathbf{w}_t$ can be represented by the encoding region R_n through the mapping $V(\mathbf{w}_t)$ expressed by

$$V(\mathbf{w}_t) = \mathbf{c}_n, \text{ if } \mathbf{w}_t \in R_n \tag{18}$$

The main idea of SLPC based vector quantization (VQ) is to find an optimal codebook such that for a given training set $\mathbf{W}$ and a codebook size N, the average distortion in representing each SLPC vector $\mathbf{w}_t$ by the closest codeword $\mathbf{c}_n$ is minimum. In mathematical terms we express

$$D^* = \min_{\mathbf{c}_n} \left[\frac{1}{T} \sum_{t=1}^{T} \min_{1 \leq n \leq N} (D(\mathbf{c}_n, \mathbf{w}_t)) \right] \tag{19}$$

where D is an LPC distortion and D^* is the average distortion of the vector quantizer.

There are several data partioning methods for the determination of an optimal VQ codebook. One of the most popular methods for VQ is the LBG algorithm [11]. The LGB-VQ method requires an initial codebook, and iteratively bi-partitions the codevectors based on the optimality criteria of nearest-neighbor and centroid conditions until the number of codevectors is reached. The spatial distortion measure we used in this study is the LPC likelihood distortion as defined in (17). The procedure for computing the codebook can be summarized as follows.

1. Given a training data set $\mathbf{W} = \{\mathbf{w}_1, \mathbf{w}_2, \ldots, \mathbf{w}_T\}$, where $\mathbf{w}_t = (w_{t1}, w_{t2}, \ldots, w_{tp})$, $t = 1, 2, \ldots, T$.
2. Given $\epsilon > 0$ (small real number)
3. Set $N = 1$, compute initial cluster center

$$\mathbf{c}_1^* = \frac{1}{T} \sum_{t=1}^{T} \mathbf{w}_t \tag{20}$$

Compute SLPC average distortion D^* using (19).
4. Splitting:

$$\mathbf{c}_{n1} = (1 + \epsilon)\mathbf{c}_n^*, \ 1 \leq n \leq N$$
$$\mathbf{c}_{n2} = (1 - \epsilon)\mathbf{c}_n^*, \ 1 \leq n \leq N$$

Set $N = 2N$
5. Set $i = 0$ and let $D^{(i)} = D^*$. Iteration:
   (a) Assign vector to closest codeword

$$V(\mathbf{w}_t) = \mathbf{c}_n^* = \arg\min_n (\|\mathbf{w}_t - \mathbf{c}_n^{(i)}\|_2)^2,$$
$$1 \leq t \leq T, 1 \leq n \leq N \tag{21}$$

   (b) Update cluster centers

$$\mathbf{c}_n^{(i+1)} = \frac{1}{|V(\mathbf{a}_t)|} \sum_{\mathbf{w}_t \in V(\mathbf{w}_t)} \mathbf{w}_t, \ 1 \leq n \leq N \tag{22}$$

   where $|V(\mathbf{w}_t)|$ is the number of $V(\mathbf{w}_t) = \mathbf{c}_n^*$.
   (c) Compute $D^{(i+1)}$ as the updated average distortion.
   (d) If

$$\frac{|D^{(i+1)} - D^{(i)}|}{D^{(i+1)}} > \epsilon \tag{23}$$

   then set $i = i + 1$, $D^* = D^{(i)}$, $\mathbf{c}_n^* = \mathbf{c}_n^{(i)}$, $1 \leq n \leq N$, and go to step (a)
6. Repeat steps 4 and 5 until the desired number of codewords is obtained.

Thus, the decision rule is made by assigning the unknown image pattern $\mathbf{w}$ to class $\omega \in \Omega$, where Ω is the set of the nuclear phases, if the spatial distortion between the unknown pattern and the codevector of the nuclear phase ω is minimum. That is

$$\omega = \arg\min_n D(\mathbf{c}_n, \mathbf{w}), \ \omega \in \Omega \tag{24}$$

4 Experimental Results

We used a dataset of HeLa cells available at Harvard Medical School, which contains 375841 cells in 892 nuclear sequences to evaluate the proposed SLPC-VQ method for classifying the nuclear phases. The average number of cells per sequence is 421. Imaging was performed by time-lapse fluorescence microscopy with a time interval of 15 minutes. Two types of sequences were used denoting drug treated and untreated. Cell cycle progress was affected by drug and some or all of the cells in the treated sequences were arrested in metaphase. Cell cycle progress in the untreated sequences was not affected. Cells without drug treatment will usually undergo one division during this period of time.

Table 1. Rates of correct classification of HeLa-cell phases

Method	Rate (%)
k-NN	82.04
k-means	85.25
VQ	85.54
SLPC-VQ	90.15

In time-lapse fluorescence microscopy images of nuclei are bright objects protruding out from a relatively uniform dark background. Thus, we carried out the segmentation by histogram thresholding. In this work the ISODATA algorithm [4] was used to perform image thresholding. By applying the ISODATA technique, each image was initially segmented into two parts using an initial threshold value. The sample mean of the gray values associated with the nuclear pixels and the sample mean of the gray values associated with the background pixels were computed. A new threshold value was then computed as the average of the two sample means. The process was repeated until the change of threshold values reaches convergence. We found that this algorithm correctly segmented most isolated nuclei, but it was not able to segment touching nuclei. The algorithm fails because it assigns the pixels to only two different groups (nuclear and background). If two nuclei are so close and there are no background pixels between them the algorithm will not be able to separate them. We therefore applied a watershed algorithm to handle this case [14].

After the segmentation process, we applied the SLPC model to extract the spatial feature of the images. We then used the VQ method to build the codebook of sizes 8, 16, and 32 for each nuclear phase model. There are 5 phases to be identified: interphase, prophase, metaphase, anaphase, and arrested metaphase. We divided the data set into 5 subsets for training 5 models and a subset for identification. Each of the 5 training sets for 5 phases contains 5000 cells, which were extracted from the cell sequences labeled from 590 to 892. These sequences were used to obtained the SLPC-VQ codebook. The identification set contains sequences labeled from 1 to 589. There are 249,547 cells in this identification set.

In our previous study [15], we used other 7 features of the nuclear images. These features include maximum intensity, mean, stand deviation, major axis, minor axis, perimeter, and compactness. Based on these 7 features, we applied the k-nearest neighbor (k-NN), k-means algorithm, and VQ to classify the five nuclear phases. The classificaton rates of the k-NN = 82.04%, k-means = 85.25%, and VQ = 85.54%. Where as the classification rates obtained from the SLPC-VQ = 88.27%, 90.32%, and 91.87% for the codebooks of 8, 16, and 32 codevectors respectively. Table 1 shows the classification rates obtained from k-NN, k-means, and VQ methods using the seven features being described above, and the average classification rate obtained the VQ method using the new spatial LPC feature.

It can be seen that various experimental results reported herein demonstrate the superiority of the proposed approach. Such new classification rate indicates a significant level of improvement over a large dataset of high-content screening. From a technical standpoint, the new result has been achieved because (1) the spatial LPC coefficients can provide a good model for extracting spatial information,(2) the distortion measure is derived based on a rigorous mathematical model for signal error comparsion, and (3) the vector quantization is an optimal computational procedure for constructing feature prototypes for pattern classification. In addition, measuring the dissimilarity between two signal patterns in terms of average or accumulated spectral distortion appears to be a very reasonable method for comparing patterns, both in terms of its mathematical tractability and its computational efficiency.

5 Conclusion

We have applied several pattern recognition methods for the classification of cell phases using time-lapse fluoresence microscopic image sequences. We have found that the new spatial LPC feature coupled with the method of vector quantization provides the best performance for the classification of cell phases. The selection of useful features is a very important task for any classifier, because effective image features can certainly enhance the performance of any classifiers. In this study we have not considered other versions of the distortion measures including the cepstral distances which have been successfully used for speech recognition [16]. The incorporations of probabilistic and fuzzy-set models to the SLPC-VQ approach may likely further improve the classification results.

Acknowledgement. The image dataset was provided by Dr. Randy King of the Department of Cell Biology, Harvard Medical School.

References

1. Giuliano, et al.: High-content screening: A new approach to easing key bottlenecks in the drug discovery process. J. Biomolecular Screening 2, 249–259 (1997)
2. Debeir, O., Ham, P.V., Kiss, R., Decaestecker, C.: Tracking of migrating cells under phase-contrast video microscopy with combined mean-shift processes. IEEE Trans. Medical Imaging 24, 697–711 (2005)
3. Fox, S.: Accommodating cells in HTS. Drug Dis. World 5, 21–30 (2003)

4. Chen, X., Zhou, X., Wong, S.T.C.: Automated segmentation, classification, and tracking of cancer cell nuclei in time-lapse microscopy. IEEE Trans. Biomedical Engineering 53, 762–766 (2006)
5. Gray, R.M.: Vector quantization. IEEE ASSP Mag. 1, 4–29 (1984)
6. Hiraoka, Y., Haraguchi, T.: Fluoresence imaging of mammalian living cells. Chromosome Res. 4, 173–176 (1996)
7. Isaaks, E.H., Srivastava, R.M.: An Introduction to Applied Geostatistics. Oxford University Press, New York (1989)
8. Itakura, F., Saito, S.: A statistical method for estimation of speech spectral density and formant frequencies. Electronics and Communications in Japan 53A, 36–43 (1970)
9. Journel, A.G., Huibregts, C.J.: Mining Geostatistics. Academic Press, London (1978)
10. Kanda, T., Sullivan, K.F., Wahl, G.M.: Histone-GFP fusion protein enables sensitive analysis of chromosome dynamics in living mammalian cells. Current Biology 8, 377–385 (1998)
11. Linde, Y., Buzo, A., Gray, R.M.: An algorithm for vector quantization. IEEE Trans. Communications 28, 84–95 (1980)
12. Makhoul, J.: Linear prediction: a tutorial review. Proc. IEEE 63, 561–580 (1975)
13. Murphy, D.B.: Fundamentals of Light Microscopy and Electronic Imaging. Wiley-Liss, Chichester (2001)
14. Norberto, M., Andres, S., Carlos, O.S., Juan, J.V., Francisco, P., Jose, M.G.: Applying watershed algorithms to the segmentation of clustered nuclei. Cytometry 28, 289–297 (1997)
15. Pham, T.D., Tran, D.T., Zhou, X., Wong, S.T.C.: Integrated algorithms for image analysis and identification of nuclear division for high-content cell-cycle screening. Int. J. Computational Intelligence and Applications 6, 21–43 (2006)
16. Rabiner, L., Juang, B.H.: Fundamentals of Speech Recognition. Prentice Hall, New Jersey (1993)
17. Yarrow, J.C., et al.: Phenotypic screening of small molecule libraries by high throughput cell imaging. Comb. Chem. High Throughput Screen 6, 279–286 (2003)

Object Detection in Watershed Partitioned Gray-Level Images

Maria Frucci and Gabriella Sanniti di Baja

Institute of Cybernetics "E.Caianiello", CNR, Pozzuoli, Italy
{m.frucci,g.sannitidibaja}@cib.na.cnr.it

Abstract. Gray-level image segmentation is the first task for any image analysis process, and is necessary to distinguish the objects of interest from the background. Segmentation is a complex task, especially when the gray-level distribution along the image is such that sets of pixels characterized by a given gray-level are interpreted by a human observer as belonging to the foreground in certain parts of the image, and to the background in other parts, depending on the local context. It very seldom happens that the background is characterized by an almost uniform gray-level. Thus, in the majority of cases, segmentation cannot be achieved by simply thresholding the image, i.e., by assigning all pixels with gray-level lower than a given threshold to the background and all remaining pixels to the foreground. One of the most often adopted segmentation techniques is based on a preliminary partition of the input gray-level image into regions, homogeneous with respect to a given property, to successively classify the obtained regions in two classes (foreground and background). In this paper, we follow this approach and present a powerful method to discriminate regions in a partition of a gray-level image obtained by using the watershed transformation. The basic idea underlying the classification is that for a wide class of gray-level images, e.g., a number of biological images, the boundary between the foreground and the background is perceived where locally maximal changes in gray-level occur through the image. Our classification procedure works well even starting from a standard watershed partition, i.e., without resorting to seed selection and region growing. However, we will also briefly discuss new criteria to be used when applying digging and flooding techniques in the framework of watershed transformation, so as to produce a less fragmented partition of the image. By using the so obtained partition of the gray-level image, the successive classification is facilitated and the quality of the obtained results is improved. Some hints regarding the use of multi-scale image representation to reduce the computational load will also be introduced.

1 Introduction

Gray-level image segmentation is a necessary step in any image analysis process to single out the subsets of the image constituting the objects of interest (foreground) and so to distinguish them from the background.

Recent surveys of different approaches to image segmentation can be found in [1,2]. Histogram thresholding (see e.g., [3]) is characterized by low computational

P. Perner and O. Salvetti (Eds.): MDA 2006/2007, LNAI 4826, pp. 94–103, 2007.
© Springer-Verlag Berlin Heidelberg 2007

complexity, but is suitable mainly for images where gray-level distribution is roughly articulated in two well defined peaks, separated by a not too broad and flat valley, i.e., images perceived as naturally binary such as written documents. To overcome these limits, rule-based methods combined with learning methods such as case-based reasoning have been developed [4]. Based on a rule set the histogram is properly smoothed and the right number of peaks is selected. Case-based reasoning ensures the incremental learning of the rule set with the proper parameters. Another approach is based on feature space clustering (see e.g., [5]), which is based on the assumption that each region of the image constitutes an individual cluster in the feature space. This method is easy to implement, but the selection of the proper features is critical and, analogously to histogram-based techniques, it does not take into account spatial information. Thus, this technique fails in presence of regions that a human observer assigns to either the foreground or the background depending on the local context. Region-based approaches (see e.g., [6]) require a suitable selection of seeds from which a growing process is done to group pixels in homogeneous regions. Of course, the selection of the seeds plays a key role for the quality of the obtained results and the method works well when the region homogeneity criterion can be defined in an easy manner. A related approach is based on edge detection techniques (see e.g., [7]). This approach follows the way in which a human observer perceives objects by taking into account the difference in contrast between adjacent regions. A segmentation method exploiting both the region-based approach and edge detection is based on the watershed transformation [8]. Fuzzy approaches use a membership function to represent the degree of some properties and are generally characterized by high computational cost. Neural network techniques can also be used to perform classification of regions, but the training phase is long and the results may be biased by the initialization phase.

The segmentation procedure to be adopted depends on the specific image domain. In this paper we consider the class of images where the distinction between foreground and background is based only on the analysis of gray-level information, without involving other features, such as the shape [9] expected to characterize the foreground components. In particular, we refer to images where the foreground is either consistently locally lighter (or consistently locally darker) than the background. This class includes, for example, a number of biomedical images. In the digitized version of a histological specimen, the regions of interest are characterized by a different gray-level, either because these regions actually have different intensity in the specimen, or because they are placed at a different depth in the slide and, hence, some of them result out of focus.

For the class of images considered in this paper, a segmentation method based on the use of the watershed transformation is the most suited one. Once the gray-level image has been partitioned into homogeneous regions, we classify the regions as belonging to either the foreground or the background, depending on the analysis of the locally maximal changes in gray-level between pairs of adjacent regions. Our classification procedure can be applied to the basic partition obtained by standard watershed transformation, i.e., without taking into account suitable procedures to select the significant seeds. Better results are achieved if the classification is accomplished on a more sophisticated watershed partition, e.g., the partition obtained by using the algorithm introduced in [10], which significantly reduces the excessive fragmentation of the input image into regions. An alternative way to reduce oversegmentation is to resort to multi-scale image representation. When a gray-level image is observed at different resolutions, only the most significant regions are perceived at all resolutions. In turn, regions with lower significance, which can be interpreted as fine details, are perceived only at sufficiently high resolution. Thus, if the seeds for watershed segmentation are detected at lower resolution, and these seeds are used to discriminate between significant and non-significant seeds in the image at full resolution, the partition is expected to consist mainly of the most significant regions.

This paper is organized as follows. In Section 2, we briefly discuss the standard watershed transformation as well as the method [10] to partition a gray-level image into a set of regions. In Section 3, we describe the procedure to classify the obtained regions in the two classes (foreground and background). In Section 4, we give some hints regarding the use of multi-scale image representation to reduce the computational load of segmentation. Finally, concluding remarks are given in Section 5.

2 Watershed Partition

The 2D gray-level input image, used in this paper as running example, has been provided by courtesy of Dr. V. Guglielmotti and includes pyramidal neurons of rabbit cerebral cortex. See Fig.1, left. Gray-levels are in the range [0, 255]. In the running example, the foreground is perceived as locally darker with respect to the background. Thus, the foreground consists of the pixels having locally lower gray-level, according to the generally followed criterion for which the highest gray-level 255 corresponds to white, while the smallest possible value 0 corresponds to black.

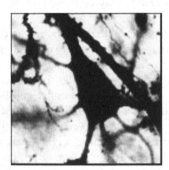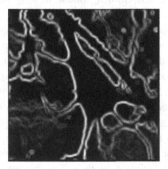

Fig. 1. The input image used as running example, left, and the relative gradient image, right

A gray-level image can be interpreted as a 3D landscape, where for every pixel in position (x,y), its gray-level plays the role of the z-coordinate. High gray-levels are mapped into mountains of the landscape, and low gray-levels into valleys. An easy way to explain how watershed transformation produces a partition of the image is the following. Let us assume that the landscape is immersed in water, after the bottom of each valley has been pierced. As a result, the valleys are flooded. Filling of a valley begins as soon as the water level reaches the bottom of that valley. A dam is built to prevent water to spread from a catchment's basin into the neighboring ones, wherever waters from different basins are going to meet. When the whole landscape has been covered by water, the basins are interpreted as the parts into which the landscape is partitioned by means of watershed lines.

In a standard watershed transformation, the bottoms of all the valleys, i.e., the regional minima, are detected in the gradient image of the input gray-level image, see Fig. 1 right. The regional minima are used as seeds for region growing. Watershed transformation generates a partition of the (gradient) image into regions characterized by homogeneity in gray-level.

As it can be seen with reference to Fig.2 left, where the watershed lines are superimposed onto the input image, the image is fragmented in a quite large number or regions (1010 for the running example). Oversegmentation is caused by the too many detected regional minima, which are not all perceptually significant.

To reduce oversegmentation, a careful selection of the regional minima to be used for region growing is necessary. Flooding and digging techniques are generally employed to cause disappearance of those regional minima that are recognized in the gradient image as corresponding to non-significant regions. Of course, the definition of significant region is crucial to obtain a meaningful partition. In [10], a new criterion has been introduced to evaluate the significance of the regions and to merge non-significant regions only with selected adjacent regions. Merging is obtained by applying again the watershed transformation on a suitably modified gradient image, which includes a smaller number of regional minima with respect to the original landscape.

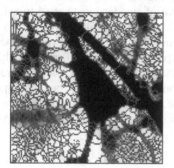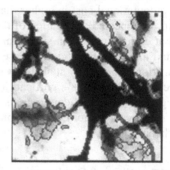

Fig. 2. Standard watershed partition, left, and watershed partition by the algorithm [10], right

In [10], as soon as the watershed partition is available, the significance of a region X is defined by evaluating the interaction of X with every adjacent region Y. Two parameters are used to define the interaction: i) the maximal depth of X when the water reaches the *local overflow pixel*, i.e., the pixel with minimal height along the watershed line separating X from Y, and ii) the absolute value of the difference in height between the regional minima of X and Y. If non-significant regions exist in the current watershed transform, the watershed transformation needs to be applied again after the seeds corresponding to the non-significant regions have been suitably removed. Three cases are possible:

1. X is significant with respect to each adjacent region Y. Then, X is definitely meaningful and no merging is necessary.

2. X is non-significant with respect to each adjacent region Y. Then, X has to be absorbed by (some) adjacent region(s). To this aim, the regional minimum of X has to be removed before applying again the watershed transformation. Flooding is accomplished by setting all pixels of X with gray-level lower than the lowest local overflow value q, to value q. X will result as merged to Y, when the watershed transformation is applied again.

3. X is significant in correspondence of some adjacent regions only. Then, X has to be merged with proper regions, selected among those with respect to which X is non-significant. Along the watershed line between X and any such a candidate region Y, a local overflow pixel exists, which is not necessarily the lowest local overflow pixel. Digging is performed, in this case, to open a canal connecting X with the region Y, which will absorb X when the watershed transformation is newly applied. The canal is identified as the minimal length path linking the regional minima of X and Y, and passing through the local overflow pixel common to X and Y. The gray-level of all the pixels in the path is set to the lower value between those of the regional minima of X and Y. When the watershed transformation is newly applied, the water can flow through the canal from X to Y, and the desired merging is obtained. The watershed lines of X, which were already detected as separating significant regions, are not altered.

The process is iterated until all resulting regions are significant. A remarkable reduction in the number of seeds, hence of the regions of the partition, is obtained. In turn, the computational cost is higher than that of standard watershed transformation, due both to the repeated application of the watershed transformation, and to the process aimed at computing region significance and possibly perform region merging via flooding and digging. For the running example, instead of 1010 regions characterizing the partition obtained by standard watershed transformation, only 259 regions are found. See Fig.2 right. Non-significant regions have been absorbed by adjacent significant regions. Non-significant regions have never been grouped to form a new, unexpected, significant region, or a region whose shape is altered with respect to the foreseen shape.

3 Classification of Regions

The watershed transformation has partitioned the image into N regions, whose membership to either the foreground or the background has not yet been established. Since the pixels constituting a region R_i of the partition don't have all the same gray-level, we compute the average, r_i, of the gray-levels of all pixels in R_i, and use it as the representative gray-level for the whole region. Adjacent regions with the same value of r_i are interpreted as constituting a single region.

We first classify the regions whose representative gray-level is smaller (greater) than the representative gray-levels of all their adjacent regions, as belonging to the foreground (background). This initial classification is done by means of a global process, which detects, in a parallel way, all gray-level local minima and local maxima. Obviously, only the pits of the valleys and the peaks of the mountains in the landscape are classified by this process.

The still unclassified regions constitute the slopes in between peaks and pits. For these regions, our classification method is inspired by visual perception and is based on the difference in gray-level between adjacent regions. In fact, the boundary separating the foreground from the background is perceived as placed wherever strong differences in gray-level occur. Thus, for any pair of adjacent regions R_i and R_j, out of which at least one is still unclassified, we compute the difference $D_{i,j}=|r_i-r_j|$. Without losing generality, we assume that the first region, R_i, in any such a pair (R_i, R_j) is the darker one and the second region, R_j, is the lighter one, i.e., we assume $r_i<r_j$.

An iterative classification process is performed, at each iteration of which the current value $\Delta=\max\{D_{i,j}\}$, i.e., the currently maximal difference in gray-level, is used to select the pairs of regions in between which the boundary is more likely to be placed. The process is iterated until all regions are classified.

At each iteration, two cases are possible, depending on the number k of adjacent regions R^k_i and R^k_j with difference Δ that are found.

When k=1, we classify the darker region R^k_i of the unique selected pair, as belonging to the foreground, and the lighter region R^k_j as belonging to the background. Moreover, we also classify in a global way all the unclassified regions that are characterized by representative gray-level not larger than r^k_i (and, hence, darker than R^k_i) as belonging to the foreground, and all unclassified regions with representative gray-level not smaller than r^k_j (and, hence, lighter than R^k_j) as belonging to the background.

When k>1, classification is still done by using a global process only if the k darker regions R^k_i have their representative gray-levels smaller than the representative gray-levels of all the lighter regions R^k_j. In other words, if the value $max_{min}= \max_k\{r^k_i\}$ is smaller than the value $min_{max}= \min_k\{r^k_j\}$, we classify all regions with representative gray-level not greater than max_{min} as belonging to the foreground, and all regions with representative gray-level not smaller than min_{max} as belonging to the background.

In turn, if at least one of the k, k>1, darker regions R^k_i has representative gray-level not smaller than the representative gray-levels of all the lighter regions R^k_j, the same global classification would lead to conflictual assignments. For example, the region with representative gray-level min_{max} should be assigned to the background, since that region is the lighter one in the pair including it, but it should be assigned to the foreground, since it results to be darker than the region with representative gray-level max_{min}. To avoid conflicts, we classify globally only the regions with representative gray-level not larger than $min_{min}=\min_k\{r^k_i\}$ (not smaller than $max_{max}=\max_k\{r^k_j\}$) as belonging to the foreground (background). For any remaining region R^k_i belonging to a pair of regions with difference Δ, the following local investigation is done. All ascending paths, consisting of unclassified regions with increasing representative gray-levels, are traced along the slope including R^k_i until a classified region is met. Since along the slope, more than one pair of adjacent regions with difference Δ can be found, a decision has to be taken to select, among the encountered pairs, the pair where the separation between the foreground and the background has to be placed. We select the pair for which r^k_i is the greatest one, so as to favor assignment of most of the slope to the foreground.

Once all regions have been classified, a final local process is accomplished, aimed at possibly changing the classification status of some regions that have been classified as belonging to the background during the iterative classification process, and are placed at the border with respect to foreground components along the slopes. This final process depends on the problem domain. If the purpose is to favor region growing without merging already detected foreground components, the change of status is done only if it does not cause a topology change. In turn, if clusters of foreground components are desired, e.g., to analyze the spatial organization of the foreground, the change of status is done only if it causes a topology change.

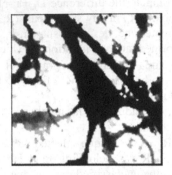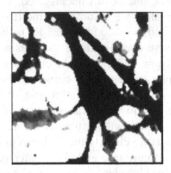

Fig. 3. Pixels classified as belonging to the foreground, starting from the standard watershed partition, left, and from the partition obtained by using the algorithm [10], right

In Fig.3, the result of the classification process is shown for the running example, starting from the standard watershed partition, left, and from the more sophisticated watershed partition [10]. Foreground pixels are shown with their original gray-levels, while all background pixels have been set to 255. Both results can be regarded as satisfactory, even if an obviously more accurate segmentation is obtained by using the partition obtained by the method [10]. As already pointed out in Section 2, the method [10] is computationally more expensive. Thus, the choice of which partition to adopt depends on a compromise between quality of the results and cost of the process.

4   Reducing Oversegmentation by Multi-scale Representation

We describe here an alternative way to reduce oversegmentation, based on the use of a multi-scale image representation. This method requires that the standard watershed transform be computed only twice, while it has to be computed for a larger number of times if the algorithm [10] is used. When observing a gray-level image at different scales, the most significant regions are perceived at all resolutions, while regions with lower significance, e.g., fine details, are perceived only at sufficiently high resolution. Thus, if the seeds for watershed segmentation of the gray-level image are detected in a representation of the image at a lower resolution with respect to the full resolution of the input image and are, then, used to distinguish significant and non-significant seeds in the image at full resolution, the resulting partition is expected to consist mainly of the regions that are perceived as the most significant ones.

To build a lower resolution representation of the input image I, we superimpose onto I a partition grid, each cell of which includes a fixed size block of pixels (*children*). We associate to each cell of the grid a single pixel (*parent*) in the representation of I at lower resolution, I'. The gray-level of a parent pixel is computed in terms of the gray-levels of its children. Depending on the position of the grid, the size of the cells, and the rule used to compute the gray-level of parent pixels, different lower resolution representations can be obtained. We here use the grid introduced in [11], whose cells are blocks of 2×2 pixels, so that the size of I' is a fourth of the size of I. The rule we adopt to compute the gray-level of the parent pixels is such to produce an almost shift invariant lower resolution image representation. Moreover, the parent-child relations are preserved, so that it is easy to transfer onto the full resolution image I, the information derived by analyzing its lower resolution representation I'.

More in detail, we inspect in forward raster fashion only pixels belonging to even rows and columns of I. This means that we use the bottom right child pixel in the 2×2 block to find the coordinates of its parent pixel in I'. For each inspected pixel in position (i,j) of I, the parent pixel in I' will be in position (i/2,j/2).

As for the gray-level of the parent pixel in position (i/2,j/2) of I', we note that the sampling grid could be placed on I in four different ways and, hence, any pixel in the 3×3 window centered on (i,j) could be the bottom right pixel of a block of the partition grid. If we consider the nine 3×3 windows, that in I are respectively centered on (i,j) and on each of its eight neighbors, then the pixel in position (i,j) is included in all the nine windows, its edge-neighbors are included in six windows and its vertex-neighbors in four windows.

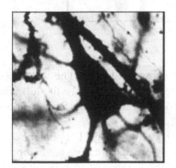

Fig. 4. Full resolution image, left, lower resolution image, middle, and markers, right

We use the above numbers 9 for the pixel in (i,j), 6 for its edge-neighbors, and 4 for its vertex-neighbors, as the proper weights to be used in a multiplicative mask to compute the gray-level of the parent pixel (i/2,j/2) of I'. By using the mask, we take into account the gray-levels of the pixel (i,j) and of its eight neighbors in a manner independent of the position of the grid. Rescaling of the computed gray-levels is done to have them still in the range [0, 255].

In Fig. 4 middle, the representation of the running example at lower resolution is shown. The full resolution image is given to the left, for the reader's convenience.

Since I is well represented by its lower resolution representation I', we can use the seeds detected in the gradient image of I', ∇', as markers to select the significant seeds in the gradient image of I, ∇. Due to the preservation of the parent-child relations, we can easily project the seeds found in ∇' onto a full resolution image. Since any parent pixel has four children, for each seed found in ∇' we identify a projected seed consisting of the union of 2×2 blocks of pixels in ∇. See Fig. 4 right.

We regard a seed detected in ∇ as significant, if the partition region associated with it in the standard watershed transform includes at least one pixel of a projected seed. Seeds originally detected in ∇, but such that the associated partition regions of the standard watershed transform do not include any pixel of projected seeds are regarded as non-significant. By means of a *flooding* process, the partition regions of the standard watershed transform corresponding to non-significant seeds are merged to adjacent regions. In practice, the gray-level of the non-significant seeds is suitably increased, so that those pixels will not be newly identified as regional minima, when the watershed transformation is applied for the second time to obtain the final partition.

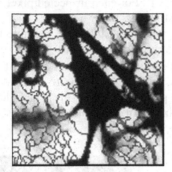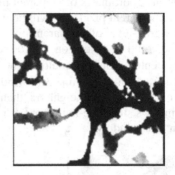

Fig. 5. Final result of the process to identify foreground components

The results of using multi-scale representation are shown in Fig. 5. The watershed lines partitioning the running example into only 153 regions are shown to the left, and the foreground components detected by using the process described in Section 3 are shown to the right, superimposed onto a uniform background.

For completeness, we point out that the resolution of the image I' could be furthermore reduced by applying to I' the same decimation process that we have applied to I. By using an image I' with even lower resolution, the number of significant seeds detected as significant in the full resolution image is expected to diminish. However, this could produce a too rough segmentation of the input gray-level image.

5 Conclusion

We have introduced a segmentation method based on a preliminary partition of a gray-level image into regions by means of the watershed transformation. The partition regions have been classified in two classes (foreground and background) by taking into account only gray-level information. Our segmentation method has been tested on a variety of images in different domains even if, in this paper, only one running example relative to biological images has been shown. The method is suited to gray-level images, where the boundary between foreground and background is perceived in correspondence with the locally maximal changes in gray-level through the image.

We have shown the classification results obtained starting from standard watershed transform, i.e., without resorting to seed selection and region growing. When this is done, the computational burden of the whole segmentation process is rather limited.

Better results are obtained, at the expense of a higher computational cost, if the gray-level image is partitioned by using a more sophisticated watershed transformation, including digging and flooding techniques to produce a less fragmented partition of the image. This more complex procedure is necessary when a finer segmentation is indispensable. We have also suggested an alternative way to reduce oversegmentation, by using multi-scale image representation. A lower resolution representation of the input image is built and the seeds for watershed partition found in this image are used as markers to discriminate between significant and non-significant seeds in the full resolution image. Segmentation done by using this approach has a cost that is intermediate between the cost of segmentation based on the standard or a more sophisticated watershed transformation, still producing good results.

Acknowledgments

We gratefully acknowledge Dr. Vittorio Guglielmotti (Neuroanatomy Research Group, Institute of Cybernetics, CNR, Pozzuoli, Naples, Italy) for kindly providing the images that we used as input to test our procedure.

References

1. Cheng, H.D., Jiang, X.H., Sun, Y., Wang, J.: Color image segmentation: advances and prospects. Pattern Recognition 34, 2259–2281 (2001)
2. Pham, D.L., Xu, C., Prince, J.L.: Current methods in medical image segmentation. Annual Review of Biomedical Engineering 2, 315–337 (2000)
3. Sahoo, P.K., Soltani, S., Wong, A.K.C., Chen, Y.C.: A survey of thresholding techniques. Comput. Vis. Graph. Im. Proc. 41, 233–260 (1988)
4. Perner, P.: An architecture for a CBR image segmentation system. Journal of Engineering Application in Artificial Intelligence, Engineering Applications of Artificial Intelligence 12(6), 749–759 (1999)
5. Bezdek, J.C., Hall, L.A., Clarke, L.P.: Review of MR image segmentation techniques using pattern recognition. Med. Phys. 20, 1033–1048 (1993)
6. Haralick, R.M., Shapiro, L.G.: Image segmentation techniques. Comput. Vis. Graph. Im. Proc. 29, 100–132 (1985)
7. Gonzalez, R.C., Wintz, P.: Digital Image Processing. Addison-Wesley, Reading, Massachusetts (1987)
8. Beucher, S., Meyer, F.: The morphological approach of segmentation: the watershed transformation. In: Dougherty, E. (ed.) Mathematical Morphology in Image Processing, pp. 433–481. Marcel Dekker, New York (1993)
9. Perner, P., Bühring, A.: Case-based object recognition. In: Funk, P., González Calero, P.A. (eds.) ECCBR 2004. LNCS (LNAI), vol. 3155, pp. 375–388. Springer, Heidelberg (2004)
10. Frucci, M.: Oversegmentation reduction by flooding regions and digging watershed lines. International Journal of Pattern Recognition and Artificial Intelligence 20(1), 15–38 (2006)
11. Frucci, M., Ramella, G., di Baja, G.S.: Oversegmentation reduction via multiresolution image representation. In: Sanfeliu, A., Cortés, M.L. (eds.) CIARP 2005. LNCS, vol. 3773, pp. 989–996. Springer, Heidelberg (2005)

Finding Cells, Finding Molecules, Finding Patterns

Carolina Wählby[1,2], Patrick Karlsson[1], Sara Henriksson[2], Chatarina Larsson[2], Mats Nilsson[2], and Ewert Bengtsson[1]

[1] Centre for Image Analysis, Uppsala University, Sweden
carolina@cb.uu.se
http://www.cb.uu.se
[2] Dept. of Genetics and Pathology, Uppsala University, Sweden
http://www.genpat.uu.se

Abstract. Many modern molecular labeling techniques result in bright point signals. Signals from molecules that are detected directly inside a cell can be captured by fluorescence microscopy. Signals representing different types of molecules may be randomly distributed in the cells or show systematic patterns indicating that the corresponding molecules have specific, non-random localizations and functions in the cell. Assessing this information requires high speed robust image segmentation followed by signal detection, and finally pattern analysis. We present and discuss this type of methods and show an example of how the distribution of different variants of mitochondrial DNA can be analyzed.

1 Introduction

Data mining can be defined as the science of extracting useful information from large data sets. In this case, the input data is digital images of cells captured using fluorescence microscopy, and the information we aim to retrieve is that of spatial distribution patterns of different variants of fluorescence labeled molecular targets. New probing and staining techniques allow a large variety of molecular targets to be visualized in situ and imaged by fluorescence microscopy. Biological processes can be studied at the ultimate level of single molecules, and with sufficient precision to distinguish even closely similar variants of molecules. It is thus possible to study the inter- or subcellular context of molecules that otherwise may go undetected at the level of populations of molecules and cells. At the same time, large numbers of cells have to be analyzed to retrieve statistically significant information. Extracting information from the resulting image data will therefore require efficient and robust cell segmentation as well as signal detection and, finally, pattern analysis.

Before signals can be assigned as coming from a particular cell, each cell has to be delineated. Segmentation is the process in which an image is divided into its constituent objects, or parts, and background. Cells can be visualized in many different ways, using different kinds of probes or stains that bind to structures within a cell. It is therefore difficult to define a single algorithm that will always

P. Perner and O. Salvetti (Eds.): MDA 2006/2007, LNAI 4826, pp. 104–114, 2007.

find the individual cells in an image, independent of method for visualization and cell morphology. Instead, cell image segmentation can be seen as a modeling problem where different approaches more or less explicitly are based on models of the cells. For example, thresholding methods can be seen as being based on a model stating that cells have an intensity that is different from the surroundings. More robust segmentation can be obtained if a combination of features, such as intensity, edge gradients, and cellular shape, is used.

In many applications in cell biology, where fluorescence marked probes are applied, the resulting images are composed of signals seen as spots of different shapes and intensities. The localization of these regions can yield important biological information. In multiple labeling experiments in particular, measurements of relative positions of regions labeled with different marker molecules can provide insight in the functional relationship between organelles and/or processes. Visual inspection is, apart from being tedious, beset with various sources of error. The positions of signals in an image should be determined automatically to derive objective information and allow further extraction of image information, such as signal intensity distribution, relative positioning and pattern analysis. The human mind is exceptional at finding patterns, it will even find patterns in data that is completely random. It is therefore valuable to have computerized methods that can search for patterns in a more objective way.

We present an image based data mining example where the distribution of different variants of the genetic information contained in mitochondria (i.e., mtDNA) has been examined. MtDNA is present in multiple copies in the mitochondrion of the cell. It is inherited together with the cytoplasm during cell replication. Genetic diseases are often caused by mutations where one single nucleotide has been substituted by another, a so-called point mutation. To be able to study and diagnose such disease with limited material from patients, there is a need for methods to detect point mutations in situ. Padlock probes and rolling circle amplification (RCA) combines highly specific target sequence recognition with a high signal-to-noise ratio. Padlock probes have been successfully used for detecting point mutations in mitochondrial DNA by [10]. We combine cell segmentation, padlock probes, signal detection and pattern analysis to examine the distribution of mtDNAs. This type of methods could also be used in applications ranging from detection of infectious organisms to studies of tumors.

2   Methods

The methods section is divided into three parts, describing methods for segmentation of cells, detection of signals, and evaluation of patterns in the detected signal distribution. A specific example is thereafter brought up in the Experiments and results section.

2.1   Cell Segmentation: Finding Cells

The difficulty of the segmentation problem is highly dependent of the type of specimen that is to be analyzed, and the result of the segmentation usually

determines eventual success of the final analysis. If we are dealing with cytological specimens where the cells are lying singly on a clean background with well
stained nuclei, and if the analysis task is limited to nuclear properties, then a
simple automatic thresholding method may be sufficient. Thresholding is often
based on histogram characteristics of the pixel intensities of the image, see [21].
In order to get a satisfactory segmentation result by thresholding, a sufficiently
uniform background is required. The transition between object and background
may be diffuse, making an optimal threshold level difficult to find also after
background correction. At the same time, a small change in the threshold level
may have a great impact on the further analysis; feature measures such as area
and volume are directly dependent on the threshold. Adaptive thresholding, i.e.,
local automatic thresholding, can be used to circumvent the problem of varying
background or as a refinement to a coarse global threshold, see [17]. The problems of segmenting clustered objects and choosing a suitable threshold level for
objects with unsharp edges will, however, remain.

If we model the objects as consisting of connected regions of similar pixels
we obtain region growing methods. A popular region growing method which
has proved to be very useful in many areas of image segmentation is the so
called watershed algorithm. The method was originally suggested by Digabel
and Lantuéjoul, and extended to a more general framework by [3]. Watershed
segmentation has then been refined and used in very many situations, see [14]
and [22] for an overview. If the intensity of the image is interpreted as elevation in
a landscape, the watershed algorithm will split the image into regions similar to
the drainage regions of this landscape. To avoid over-segmentation, i.e., splitting
of the image into too many regions, water can be allowed to rise only from places
marked as seeds [2,9,11,14,22]. Seeds may be found manually or by automated
methods. Over-segmentation can also be reduced by rule-based merging, e.g.,
shown by [15].

Cell nuclei are usually convex and fairly round or elliptic and the shape can
therefore be used as part of the object model. Touching nuclei that are not
separated by an intensity threshold can be separated by distance transforming
[4] the binary image and applying watershed segmentation, see work done by
[12], [17], and [23].

None of the above described methods will alone produce a satisfactory result
on the more difficult types of cell and tissue images. We may for instance have
problems if (1) the cells are clustered, (2) the image background is varying, and
(3) there are intensity variations within the cells. By combining the methods,
more powerful models can be created, and more complex segmentation problems
be solved. Our experience is that the seeded watershed approach is a useful
core component in such segmentation models. Complex segmentation methods
often require a large number of input parameters that have to be optimized
for each type of input data. In case based reasoning, the segmentation step is
initiated by classifying each image as belonging to one of a number of pre-defined
cases, and input parameters optimized for the particular case are applied during
segmentation, see [19].

2.2   Signal Detection: Finding Molecules

The most common method for finding structures such as proteins and organelles in situ is using antibodies labeled with fluorescing molecules. Fluorescence labeled secondary antibodies can be used to amplify the signal and increase signal to noise ratios. The genetic information contained in the DNA in a cell can be stained as a whole using non-specific chemical dyes, or in a more specific way using oligonucleotide probes that search for a particular DNA sequence. Fluorescence in situ hybridization (FISH) is such a method, and can detect larger mutations such as duplications, translocations and deletions, but it is not sensitive enough to distinguish between single nucleotide sequence variations. Primed in situ labeling (PRINS) reaction uses a specific primer that will initiate synthesis of DNA from fluorescent labeled nucleotides at the site of sequence detection, see [7]. The method does however not give signals from single-copy genes that are distinguishable from noise caused by insertions of fluorescing nucleotides in other places in the genome. In the oligonucleotide ligation assay (OLA), as shown by [8], oligonucleotides are hybridized juxtaposed with the junction at the point mutation. If there is a perfect match, the two probes can be enzymatically hybridized and detected. There is however a risk of wrong probes being ligated, especially when trying to find many different sequence variants in the same sample. This can be avoided by, instead of using two separate probes, using a single linear probe, a so called padlock probe. The padlock probe has ends that are designed to hybridize juxtaposed at the point mutation and if correctly base paired at the point mutation, the two ends can be enzymatically ligated, forming a circular DNA molecule, see [16]. The specifically reacted circular DNA can thereafter be amplified using rolling circle amplification (RCA) generating molecules that are bound by hundreds of fluorescing probes, see [1]. These signals can be detected by fluorescence microscopy as bright spots at or below the resolution of the microscope, the image resolution limited by the point spread function of the microscope.

An image that contains multiple, and sometimes clustered spots with different maximum intensities can be segmented in many different ways. Regions found by procedures such as intensity thresholding often contain more than one local maximum of intensity, indicating that the region consists of more than one spot. Top-hat transforms, see [5], in combination with threshold procedures fail to divide the image into separate domains each containing one local maximum of intensity as the top-hat transform is unable to distinguish a local maximum from a saddle-point. If each spot contains a single local maximum, watershed segmentation, as described above, in combination with a background threshold, may be used to delineate individual signals. Another approach is the largest contour segmentation by [13], where the domain of each signal is defined by a local maximum and an iterative region-growing. If two or more signals are clustered into a spatially large signal, where the individual signals do not contain individual intensity maxima (due to tight clustering or signal saturation), the shape of the signal can provide clues as to how the signals should be detected. In work by [6], the curvature of the edge of each signal cluster is examined, and signals are positioned within the cluster starting from the position where the greatest curvature is found.

2.3   Hypotheses Testing: Finding Patterns

Patterns in image data can be evaluated by interpreting the signal distribution as image texture, and using texture measurements. Some of the most commonly used texture measures are derived from the Grey Level Co-occurrence Matrix (GLCM). The GLCM is a tabulation of how often different combinations of pixel brightness values (grey levels) occur in a pixel pair in an image, see [5]. Multi-interval discretization has also shown to give useful information for cell pattern classification, see [20]. Different kinds of distance measures can also be used to evaluate spatial relationships between signals once they have been detected. In the case where we want to know if red and green signals are randomly distributed in the cytoplasm or not, we can simply count how often a red signal has a green signal as its closest neighbor, and how often a red signal has a red signal as its closest neighbor (and the other way around). To evaluate the outcome, we have to know what distributions can be expected. By creating a virtual cell, where possible positions and number of signals of different types is given as input, different hypotheses can be tested. We can then compare the spatial relationships between signals in real cells with those in a virtual cell with the same input parameters. Signals in the virtual cell can be positioned based on a hypotheses, i.e., either randomly, or according to a pattern. Thousands of randomized virtual cells can thereafter be created, and the probability of the real cell having the hypothesized signal distribution pattern can be examined. Factors such as staining efficiency and noise may also be added to the virtual cell for comparison.

3   Experiments and Results

To illustrate the concepts discussed we will here describe a project where model based cell segmentation is combined with padlock probing for molecule detection, model based signal detection, and pattern analysis to examine the spatial distribution of mtDNAs.

3.1   Finding Cells

In the presented experiment, no general stain defining the cytoplasm is available. We do however have a general stain defining the nucleus of each cell. Combining this information with the fact that the over all signal variance is higher within the cytoplasm than in the image background, a model defining cytoplasms is created. The three markers (i.e. nuclear stain, padlock probe 1, and padlock probe 2) are shown as three images, see Fig. 1 A, B, and C, each captured with a different filter set in the fluorescence microscope. Cells segmentation is initiated by intensity thresholding of the image showing the nuclear stain.

A suitable threshold is found using Otsu's method, which searches for the threshold level that minimizes the intra-class variance of foreground as well as background, see [18]. The resulting binary image is shown in Fig. 1 D. Intensity thresholding is not enough to separate nuclei that are very close to one another. Thanks to their round shape, touching nuclei can be separated by applying watershed segmentation to a distance map of the binary image. Over-segmentation due to multiple local maxima is avoided by smoothing of the distance map. The distance map is shown in Fig. 1 E, and the result after watershed segmentation is shown in Fig. 1 F. The region surrounding each nucleus, not belonging to the image background, is the cytoplasm. The image background has less variance than the parts of the image containing cells, and can thus be found by variance filtering. The variance map of the sum of B and C (Fig. 1 G) is thresholded using Otsu's method. Small, disconnected regions are removed by morphological opening, and larger disconnected regions are re-connected by dilation. In both cases, a disc-shaped structure element of radius 10 pixels was used. The choice of structure element depends on size and density of signals.

Each pixel of the resulting binary image (Fig. 1 H) is assigned to the closest nucleus by seeded watershed segmentation, using the segmentation result from the segmentation of the nuclei as seeds. Non-seeded regions are discarded as background. Fig. 1 I shows the final segmentation result on top of a projection of the three input images.

3.2   Finding Molecules

In the presented experiment, a model system with padlock probes was used. It consists of different detection sequences that represent real point mutations. Four different padlock probes were used for testing efficiency of staining and evaluation of signal distribution patterns. Two of the padlock probes hybridize to different sites on the same mtDNA fragment, i.e., they are non-competing. One is detected using Cy3 (red), and one using FITC (green). The other two probes bind to the same site, and are therefore competing.

Signal detection was initiated by first reducing the background variation present in the images. As the cells are cultured on a glass surface, they are comparably flat. Despite this, it is necessary to image them in more than one focal plane to make sure that all signals are detected. In the presented study, the slides were studied in a fluorescence microscope (Axioplan II Zeiss) using a 63x objective. Images were collected with Axiovision 4.3 software as a 16 layers z-stack with 0.5um between consecutive layers. The nuclear stain DAPI emission was collected at 360nm excitation wavelength for 200ms, green padlock detection fluorochrome FITC at 470nm for 200ms and red padlock detection fluorochrome Cy3 at 546nm for 450ms. Background was reduced in each z-image separately by morphological tophat filtering, using a disc of radius 10 pixels. Tophat filtering removes intensity variations that have a spatial extent greater than that of the disc. The 16 layers were thereafter combined using maximum intensity projection. Projection of the 3D information to a single image will result in loss

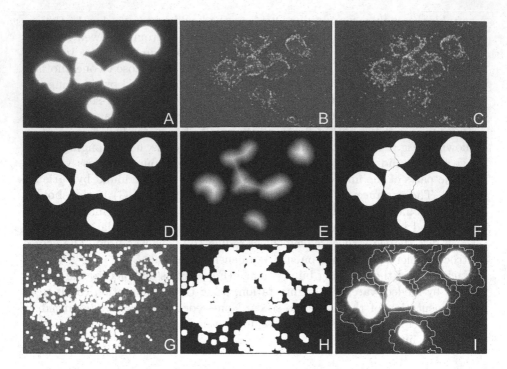

Fig. 1. A: Part of an image of DAPI stained cell nuclei. **B**: Image of the same cells showing logarithm of signals from padlock probe 1 (stained with Cy3) and **C**:, logarithm of signal from padlock probe 2 (stained with FITC). **D**: Binary image after thresholding of **A**. **E** shows the same image after distance transformation, and **F** is the result after watershed segmentation on the distance map, i.e., the final segmentation of the nuclei. **G**: The cytoplasm is found by combining the two images showing padlock probes (**B** and **C**) and applying a variance filter. **H**: Potential cytoplasm after thresholding of variance map and morphological opening to remove noise. **I**: Final segmentation result based on shape of nuclei and variance of cytoplasm.

of spatial information in the z-direction. As the extension of cultured cells in z-direction is only a fraction of their extension i x-, and y- direction, the 3D information was considered less important. This would however not be the case if cells in tissue were observed.

The result after pre-processing and maximum intensity projection of a small fraction of an image is shown in Fig. 2 A. Simple intensity thresholding will separate the signals from the image background, but signals that are clustered will not be separated from each other. To separate clustered signals, Watershed segmentation, starting from all local maxima, is applied to the image, and the watershed regions are allowed to extend until they reach a predefined background threshold. The resulting signal centers after watershed segmentation of Fig. 2 A are shown in Fig. 2 B.

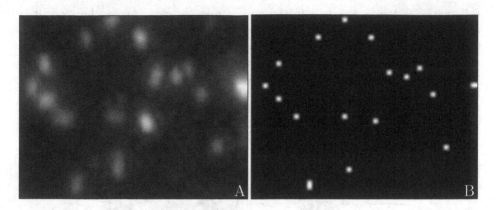

Fig. 2. A: Enlarged image showing signals from padlock probes after image pre-processing. **B**: Positions of detected signals using watershed segmentation.

3.3   Finding Patterns

The patterns of red and green signal distributions were examined by searching for aggregations of signals, i.e., the existence of groups of signals with the same color. The affinity of red and green signals was measured as the number of red signals with a green nearest neighbor, the number of red signals with a red nearest neighbor, the number of green signals with a green nearest neighbor, and finally the number of green signals with a red nearest neighbor. To normalize the observed result and evaluate the probability of non-random pattern, virtual cells with truly random patterns were created. Virtual cells with random signal distributions were created by keeping the number of red and green signals the same as in the real cell. Red and green signals were then randomized within an area corresponding to the cytoplasm region with the 10% greatest variance. Virtual cells were re-created 1000 times, neighborhood relations were examined, and the resulting distributions were compared with observed distributions.

A number of restrictions have to be taken into consideration when creating the virtual cells. First of all there is a limit in closeness between signals in the real data due to the point-spread function. Two signals that are of the same color will not be separated if they are closer than the width of a single signal. This has to be compensated for in the randomized data, or else it will affect the outcome of the analysis of the neighbor relations. Randomized signals that appeared closer to one another than the two closest signals in the real data were simply removed, and a new pair of random signals was created and tested for closeness with existing randomized signals. Fig. 3 A, top, shows the true signal distribution within the cytoplasm of a cell with competing padlock probes, red and green signals as + and o respectively. Fig. 3 A, bottom, shows one of the 1000 virtual cells with randomized signals. Fig. 3 B, top, shows the true signal distribution within the cytoplasm of a cell with non-competing padlock probes, red and green signals as + and o respectively. Fig. 3 B, bottom, shows one of the 1000 virtual cells with randomized signals in the non-competing case. As can be seen, it is not trivial to pick out the cells showing random distributions compared to those showing a non-random affinity between red and green signals. Comparing the randomized data with the true signal distributions shows that the pattern falls within the randomized distribution in the case with competing probes, while the non-competing probes show a red-green affinity three standard deviations greater than that of the randomized distribution. This agrees with what one would expect as the non-competing probes can bind to the same mtDNA fragment.

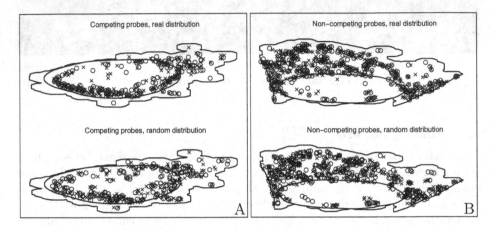

Fig. 3. Cells treated with two competing (**A**) or non-competing (**B**) padlock probes in a 50-50 concentration. Top: true signal distribution, bottom: randomized signal distribution. Red and green signals as + and o respectively. Nucleus and cytoplasm outlined.

4   Conclusions

Patterns and spatial relationships between molecules in cells are of great interest in many types of analysis. One way of examining patterns in cells is by visualizing the molecules of interest using highly specific detection probes, and comparing observed signal distributions with randomized distributions in a virtual cell. Before patterns can be examined, signals from detection probes have to be found, and clustered signals separated. More than one cell is often observed simultaneously, and automated identification of each individual cell in an image provides efficient analysis of large data sets with little impact from observer bias. In order to obtain a successful cell segmentation method it is important to use as much a priori information as possible about the appearance of the objects that are to be segmented, without resorting to models that are too complex or too difficult to train or apply.

References

1. Banér, J., Nilsson, M., Mendel-Hartvig, M., Landegren, U.: Signal amplification of padlock probes by rolling circle replication. Nucleic Acid Research 26, 5073–5078 (1998)
2. Beucher, S.: The watershed transformation applied to image segmentation. Scanning Microscopy 6, 299–314 (1992)
3. Beucher, S., Lantuéjoul, C.: Use of watersheds in contour detection. In: International Workshop on Image Processing: Real-time and Motion Detection/Estimation, Rennes, France (September 1979)
4. Borgefors, G.: Distance transformations in digital images. Computer Vision, Graphics and Image Processing 34, 344–371 (1986)
5. Haralick, R.M., Shapiro, L.G.: Computer and Robot Vision, vol. 1, 2. Addison-Wesley, Reading, Mass. (1992)
6. Karlsson, P., Lindblad, J.: Segmentation and separation of point like fluorescent markers in digital images. In: Proceedings of 2004 IEEE International Symposium on Biomedical Imaging, pp. 1291–1294. IEEE Computer Society Press, Washington D.C., USA (2004)
7. Koch, J., Kolvraa, S., Petersen, K., Gregersen, N., Bolund, L.: Oligonucleotide-priming methods for the chromosome-specific labeling of alpha satellite dna in situ. Chromosoma 98, 259–265 (1988)
8. Landegren, U., Kaiser, R., Sanders, J., Hood, L.: A ligase-mediated gene detection technique. Science 241, 1077–1080 (1988)
9. Landini, G., Othman, I.E.: Estimation of tissue layer level by sequential morphological reconstruction. Journal of Microscopy 209(2), 118–125 (2003)
10. Larsson, C., Koch, J., Nygren, A., Janssen, G., Raap, A.K., Landegren, U., Nilsson, M.: In situ genotyping individual dna molecules by target- primed rolling- circle amplification of padlock probes. Nature Methods 1, 227–232 (2004)
11. Lockett, S.J., Sudar, D., Thompson, C.T., Pinkel, D., Gray, J.W.: Efficient, interactive, and three-dimensional segmentation of cell nuclei in thick tissue sections. Cytometry 31, 275–286 (1998)
12. Malpica, N., de Solorzano, C.O., Vaquero, J.J., Santos, A., Vallcorba, I., Garcia-Sagredo, J.M., del Pozo, F.: Applying watershed algorithms to the segmentation of clustered nuclei. Cytometry 28(4), 289–297 (1997)
13. Manders, E.M.M., Hoebe, R., Strackee, J., Vossepoel, A.M., Aten, J.A.: Largest contour segmentation: a tool for the localization of spots in confocal images. Cytometry 23, 15–21 (1996)
14. Meyer, F., Beucher, S.: Morphological segmentation. Journal of Visual Communication and Image Representation 1(1), 21–46 (1990)
15. Najman, L., Schmitt, M.: Geodesic saliency of watershed contours and hierarchial segmentation. IEEE Transactions on Pattern Analysis and Machine Intelligence 18(12), 1163–1173 (1996)
16. Nilsson, M., Malmgren, H., Samiotaki, M., Kwiatkowski, M., Chowdhary, B.P., Landegren, U.: Padlock probes: Circularizing oligonucleotides for localized dna detection. Science 265, 2085–2088 (1994)
17. de Solorzano, C.O., Garcia Rodriguez, E., Jones, A., Pinkel, D., Gray, J., Sudar, D., Lockett, S.: Segmentation of confocal microscope images of cell nuclei in thick tissue sections. Journal of Microscopy 193, 212–226 (1999)
18. Otsu, N.: A threshold selection method from gray-level histograms. IEEE Trans. on System Man and Cybernetics 9(1), 62–69 (1979)

19. Perner, P.: An architecture for a CBR image segmentation system. IEEE Journal on Engineering Application in Artificial Intelligence, Engineering Applications of Artificial Intelligence 12(6), 749–759 (1999)
20. Perner, P., Perner, H., Müller, B.: An mining knowledge for HEp-2 cell image classification. Journal Artificial Intelligence in Medicine 26, 161–173 (2002)
21. Sahoo, P.K., Soltani, S., Wong, A.K.C., Chen, Y.C.: A survey of thresholding techniques. Computer Vision, Graphics and Image Processing 41, 233–260 (1988)
22. Vincent, L.: Morphological grayscale reconstruction in image analysis: Applications and efficient algorithms. IEEE Trans. on Image Processing 2(2), 176–201 (1993)
23. Wählby, C., Sintorn, I.-M., Erlandsson, F., Borgefors, G., Bengtsson, E.: Combining intensity, edge, and shape information for 2D and 3D segmentation of cell nuclei on tissue sections. Journal of Microscopy 215(1), 67–76 (2004)

Automatic Fuzzy-neural Based Segmentation of Microscopic Cell Images

Sara Colantonio[1], Igor Gurevich[2], and Ovidio Salvetti[1]

[1]Institute of Information Science and Technologies, Italian national Research Council,
via G. Moruzzi 1, 56124 Pisa, Italy
{Sara.Colantonio,Ovidio.Salvetti}@isti.cnr.it
[2] Dorodnicyn Computing Center, Russian Academy of Sciences, Vavilov str. 40,
119991 Moscow, Russian Federation
{igourevi}@ccas.ru

Abstract. In this paper, we propose a novel, completely automated method for the segmentation of lymphatic cell nuclei represented in microscopic specimen images. Actually, segmenting cell nuclei is the first, necessary step for developing an automated application for the early diagnostics of lymphatic system tumours. The proposed method follows a two-step approach to, firstly, find the nuclei and, then, to refine the segmentation by means of a neural model, able to localize the borders of each nucleus. Experimental results have shown the feasibility of the method.

Keywords: Microscopic Cell Images; Color Image Segmentation; Fuzzy Clustering; Artificial Neural Networks.

1 Introduction

A great deal of research has concerned, in the last years, the development of automated systems for the early diagnosis of lymphatic tumors based on the morphological analysis of blood cells in microscopic specimen images. Actually, pathologists usually make diagnosis by analyzing the morphology of specimen cells [1] [2].

The first and necessary step for automating cell analysis is an accurate segmentation of the cells themselves, which is then followed by the extraction of significant morphological parameters. Unfortunately, cell segmentation is usually an ill-posed problem: due to poor dye quality, cell boundary could be not well distinguishable and parts of the same tissue could be not equally stained; two or more cells could be very close to each other or even overlapping; the chromatin distribution inside the cells could generate strong computed edges which mislead the segmentation.

In past years, many segmentation methods have been presented [3] [4]. They include watersheds [5] [6], region-based [7] and threshold-based methods [8]. The limit of these methods consists of the lack of shape information of the cell, which can be useful in presence of noise.

Recently, the application of *Active Contours* has been widely investigated for cell segmentation [9] [10]. However, such methods require an initialization of the *snake*,

P. Perner and O. Salvetti (Eds.): MDA 2006/2007, LNAI 4826, pp. 115–127, 2007.

making the segmentation not completely automated. Moreover, having to select which cell the snake should be apply to, much information regarding all the cells represented in the images is lost.

Other contour-based methods include *Active Shape Models* (ASM) [11], *Active Appearance Models* (AAM) [12] and variational deformable models (*Strings*) [13]. In the first two cases, a boundary model and its allowed variations are learned from a set of example boundaries and represented by a set of labeled points, encoding only shape information in ASM, also image features in AAM. The Strings method differs from the previous ones in adopting a continuous instead of discrete boundary representation, together with a multiple features description, giving place to a multivariate curve representation in functional space (instead of a point representation in vector space). All these methods require initialization and allow modeling only the variation seen in the training set of boundary examples.

The application of case-based object recognition has been also investigated [14]: to capture the great variation in cell appearance, a set of cases is learned, stored in a case base, and then used for segmentation by means of a matching procedure based on a similarity measure.

The method we propose in this paper has the main characteristic to be completely automated. Moreover, it is suitable to segment all the cells contained in the images, thus allowing for the extraction of data not only from the malignant ones.

Following a two-step approach, images are first clustered, in order to perform a rough segmentation and localize the cells. In a second processing step, an *Artificial Neural Network* (ANN) is applied to the image portions containing the localized cell for individuating cell borders.

Such an approach assures a high level of robustness, because the ANN performs a classification of the image and then it can distinguish among different kinds of structures, e.g. cell nucleus, cytoplasm, background, artifacts and so forth.

2   The Fuzzy-neural Segmentation

Microscopic cell images are acquired as footprints of lymphoid tissue stained according to the Romanovsky-Giemsa technique and digitized as color images.

Each image I contains a number, say n, of cells which are constituted by the internal body – the nucleus –, which is the structure of interest to be segmented, and the cytoplasm. Due to the staining procedure, artifacts can be present in the images, as well as not perfectly stained cells that can be then considered as added *noise*.

The proposed method is suitable to detect nuclei borders and consists in applying to each image I a two-stage procedure as follows:

- *Cell dislocation detection*: a cluster analysis, based on the *fuzzy c-means* algorithm, is applied to identify and label homogeneous regions in the image. The clustered regions are then used to divide the entire image in disjoint sub-parts for further processing (image partition).
- *Cells contours extraction*: from each image partition relevant features are extracted and a dedicated ANN is used to complete the segmentation by identifying the contours of each cell.

A sketch of the method is shown in Fig. 1.

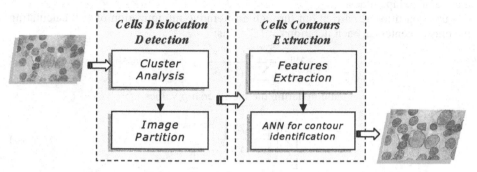

Fig. 1. The two-step method for cell segmentation

In the following, each step is described in more details.

2.1 Cells Dislocation Detection

In order to individuate how cells are dislocated in the microscopic images, a fuzzy cluster analysis is performed and each image is partitioned in disjoint parts for next step elaboration.

Cluster Analysis.   Homogeneous image regions are labeled using an unsupervised clustering method, based on the *fuzzy c-means* algorithm (FCM) [15]. This algorithm groups a set of data in a predefined number of classes so as to iteratively minimize a criterion function, namely the sum-of-squared-distance from region centroids, weighted by a cluster membership function. A membership grade $p \in [0,1]$ is associated to each element of the data set, describing its probability to belong to a particular cluster.

For each cell image I, a features vector

$$(I_0(\mathbf{x}), I_1(\mathbf{x}), I_2(\mathbf{x}),..., I_q(\mathbf{x}))$$

is computed for any pixel x, considering $I(\mathbf{x})$ as a vector of the three color component $I(\mathbf{x}) = (r,g,b)$. Then $I_0(\mathbf{x}) = I$, and for $k = 1,...,q$, $I_k(\mathbf{x}) = I*\Gamma_k(\mathbf{x})$, where Γ_k is a Gaussian filter with $\sigma = k$. In this way, we obtain a data set $D = \{\mathbf{v}_1, \mathbf{v}_2, ..., \mathbf{v}_m\}$ where each $\mathbf{v}_h$, $h=1,...,m$ is a vector in $\mathfrak{R}^p$ representing image elements at different resolutions.

Let U_{cm} be a set of real $c \times m$ matrices, with c being an integer, $2 \leq c < m$; the fuzzy c-partition space for D is, then, the set:

$$\Omega = \{U \in U_{cm} : u_{ih} \in [0,1], \sum_{i=1}^{c} u_{ih} = 1, 0 < \sum_{h=1}^{m} u_{ih} < m\} \qquad (1)$$

where u_{ih} is the membership value of $\mathbf{v}_h$ in cluster i ($i = 1,...,c$).

By applying FCM, an optimal fuzzy c-partition and corresponding prototypes are found minimizing the objective function:

$$J_\eta(U,\Lambda;D) = \sum_{h=1}^{m} \sum_{i=1}^{c} (u_{ih})^\eta \|\mathbf{v}_h - \lambda_i\|^2 \qquad (2)$$

where $\Lambda = (\lambda_1, \lambda_2,..., \lambda_c)$ is a matrix of unknown cluster centers $\lambda_i \in \mathfrak{R}^p$, $\|\cdot\|$ is any norm, e.g. the Euclidean norm, expressing the similarity between each data vector $\mathbf{v}_h$

and the center λ_i, and the weighting exponent $\eta \in [0,\infty)$ is a constant that influences the membership values.

Fuzzy partition is carried out through an iterative minimization of (2), calculating the cluster centers at each iteration $t = 1,2,\dots$ as:

$$\lambda_i^{(t)} = \frac{\sum_{h=1}^{m}(u_{ih}^{(t)})^\eta \mathbf{v}_h}{\sum_{h=1}^{m}(u_{ih}^{(t)})^\eta} \tag{3}$$

and updating the membership values as:

$$u_{ih}^{(t)} = \left[\sum_{j=1}^{c}\left(\frac{\left\|\mathbf{v}_h - \lambda_i^{(t)}\right\|^2}{\left\|\mathbf{v}_h - \lambda_j^{(t)}\right\|^2}\right)^{\frac{2}{\eta-1}}\right]^{-1} \tag{4}$$

The iterative process stops when $|U^{(t+1)}-U^{(t)}|$ follows under a certain threshold or the maximum number of iterations is reached.

Applying the FCM on the cell images induces a partition of each *slide* into a set $P=\{R_1,R_2,\dots\}$ of disjoint connected regions R, where the indices $1,2,\dots$ are region labels. In other words, by clustering, we obtain a rough segmentation which can be refined reducing the computation by the following step of image partitioning.

Image Partitioning. Once clustered the image, the convex hull of each connected region is calculated in order to delimitate the largest image portion (*convex image*) containing the corresponding connected region.

Starting from the convex hull, an image partition is extracted slightly enlarged in both directions the *convex image*. Such partition contains what the FCM has classified as a unique cell. However, the contour of the clustered region can be inaccurate, including, for instance, the cytoplasm; moreover, it can happen that two very closed or touching cells are clustered as a unique region. For these reasons, it is necessary to refine the clustering in a further step.

2.2 Cells Contour Extraction

In order to detect the exact cell contour, from each image partition, a set of features is extracted and classified by a dedicated ANN.

Features Extraction. Analyzing the properties of cell images and of the similar cells, the following vector of features $\Im(\mathbf{x})$ is computed for characterizing each pixel $\mathbf{x}$ of the segmented image partition:

- *Color values*: $I(\mathbf{x}) = (r,g,b)$;
- *Mean color value*: $M(\mathbf{x}) = (M_r, M_g, M_b)$ computed applying an average filter $F(\mathbf{x})$, i.e. $M(\mathbf{x}) = I(\mathbf{x}) *F(\mathbf{x})$;
- *Gradient norm*: $\|\nabla I(\mathbf{x})\|$ and its mean, computed along the three color components;
- *Radial gradient*: $G_{rt}(\mathbf{x})$, defined as the gradient component in the radial direction $\hat{r}$ from the centre of the connected region;
- *Radial position*: $P_r(\mathbf{x})$, computed in the radial direction;
- *Membership value to the clustered region*: $u_i(\mathbf{x})$, where i is the cluster index considered as a cell in the image partition.

ANN for contours identification. The vectors of the extracted features $\Im(\mathbf{x})$ are processed by a dedicated ANN. It consists in a *Multilayer Perceptron*, trained according to the *Error Back-Propagation* (EBP) algorithm [16] to recognize five different classes. At present, to resolve ambiguity in case of touching cells and let the network learn and generalize better, five pixel classes are selected:

  1. Cell border
  2. Cell internal body
  3. Cytoplasm
  4. Background
  5. Artifact

Let $o_j(\Im(\mathbf{x}))$ be the answer of the output units of the network when the features vector $\Im(\mathbf{x})$ is being processed; then, the pixel membership to one of the above mentioned classes can be computed as

$$\Phi(\mathbf{x}) = \text{argmax}_{j=1,\dots,5}(o_j(\Im(\mathbf{x})))\,. \tag{5}$$

A set of pre-classified images has been used to train the network, using the *Resilient Back-Propagation* [17] version of the EBP algorithm. Once defined the desired ψ_p output for each input vector of the training set $TS = \{\Im_p(\mathbf{x})\}$, the cost function

$$E = \frac{1}{2}\sum_{p=1}^{|TS|}(\psi_p - \mathbf{o}_p)^2 \tag{6}$$

where $\mathbf{o}_p = (o_1, o_2, \dots, o_j)$ is the output vector of the network, is minimized iteratively computing the weight update at each iteration step t as follows:

$$\Delta w_{ij}^{(t)} = \begin{cases} -\Delta_{ij}^{(t)} & \text{if } \dfrac{\partial E}{\partial w_{ij}}(t) > 0 \\[2mm] -\Delta_{ij}^{(t)} & \text{if } \dfrac{\partial E}{\partial w_{ij}}(t) > 0 \\[2mm] 0 & \text{otherwise} \end{cases} \tag{7}$$

where w_{ij} is the weight between the network units i and j, and Δ_{ij} is the amount of weight change which, starting from a chosen value Δ_0, varies at each step t according to the following equation:

$$\Delta_{ij}^{(t)} = \begin{cases} \varepsilon^+ \Delta_{ij}^{(t-1)} & \text{if } \dfrac{\partial E}{\partial w_{ij}}(t-1) \cdot \dfrac{\partial E}{\partial w_{ij}}(t) > 0 \\[2mm] \varepsilon^- \Delta_{ij}^{(t-1)} & \text{if } \dfrac{\partial E}{\partial w_{ij}}(t-1) \cdot \dfrac{\partial E}{\partial w_{ij}}(t) < 0 \\[2mm] \Delta_{ij}^{(t-1)} & \text{otherwise} \end{cases} \tag{8}$$

where $0 < \varepsilon^- < 1 < \varepsilon^+$ are parameters used to regulate weight modifications.

The final result of this step is discussed in the following section.

3 Results

Footprints of lymphoid tissues were Romanovsky-Giemsa stained and digitized with digital camera mounted on Leica DMRB microscope using PlanApo 100/1.3 objective. The equivalent size of a pixel was 0,0064 μ^2; 24-bit color images were stored in TIFF format of dimensions 1200 × 1792. A total number of 800 microscopic images were considered, with an average number of 20 cells for each. An example of a microscopic cell image and its three color components is reported in Fig. 2.

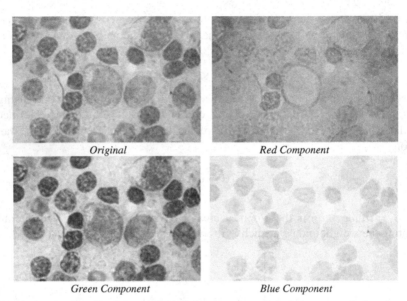

Original   *Red Component*

Green Component   *Blue Component*

Fig. 2. An example of microscopic cell image: the original image and the three color components

The cluster analysis was designed to be performed on the features vectors $(I_0(\mathbf{x}),$ $I_1(\mathbf{x}), I_2(\mathbf{x}),..., I_q(\mathbf{x}))$ with $q = 5$, but, among such components, only $I_3(\mathbf{x})$ and $I_5(\mathbf{x})$ were considered relevant. The input vectors represented in the form of color images are shown in Fig. 3. The same feature vector for each of the color components of the image is reported in Fig. 4.

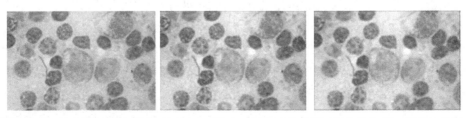

Fig. 3. An example of the three-component feature vector used for clustering: from left to right, original, $\sigma = 3$ and $\sigma = 5$

Red component

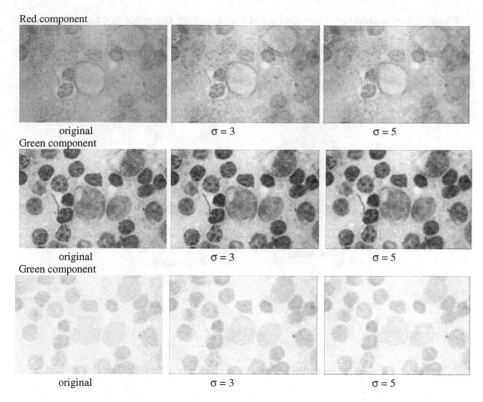

original   σ = 3   σ = 5

Fig. 4. An example of the feature vector with the original values $I_0(x)$ and $I_3(x)$ and $I_5(x)$ for each of the three color components

The FCM algorithm is applied to divide image pixels into two clusters corresponding to cell and background. A filling operation is performed to eliminate little holes, while clustered regions of negligible area are deleted. An example of the clustering results is reported in Fig. 5.

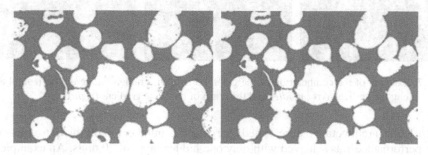

Fig. 5. Example of the clustering results: rough clustered image (*left*), clustered image after a filling operation and after deletion of regions of negligible area (*right*)

Examples of image partitions extracted for detecting the exact borders of a cell are shown in Fig. 6.

From each partition, the set of the mentioned features is extracted. To illustrate the significance of such set, Fig. 7 shows an example of the gradient regarding the green component.

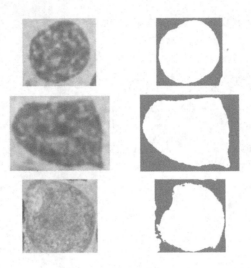

Fig. 6. Image partitions containing the cells to be segmented

The set of 800 images was partitioned in (i) a sub-set of 300 images, used for training, and (ii) a sub-set of the remaining 500 images used for the testing phase. A semi-automatic segmentation was performed for the training set, consisting in a classification of images according to the different classes of pixels.

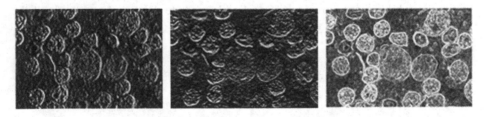

Fig. 7. Example of the computation of the green component gradient along the horizontal axis (*left*), along the vertical axis (*middle*) and the norm of the same gradient (*right*)

Different architectures were tested, varying the number of the hidden units: the best performance was achieved with only one hidden layer of 20 units. An example of the segmentation results is illustrated in Fig. 8, where the entire classification results are reported too.

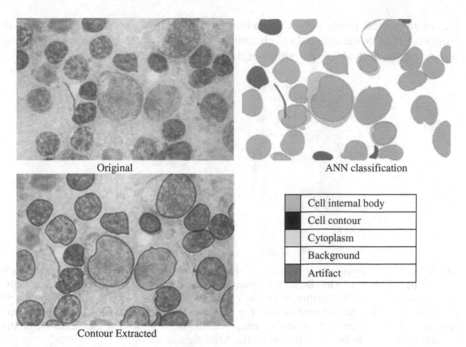

Fig. 8. Example of segmentation. *upper left*: original cell image; *upper right*: results of the ANN classification (five classes with different colors); *lower left*: identified contours of each cell; *lower right*, color legend.

4 Discussion and Conclusions

A two-step method for segmenting microscopic cell images has been presented.

The first step consists of a fuzzy clustering of images performed to obtain a rough segmentation and to detect cell dislocation. In the second step, a dedicated ANN is applied to refine the segmentation by discriminating image components, i.e. cell borders, cell internal body, cytoplasm, background, and artifacts.

The main features of the proposed method are
- complete automation of segmentation
- possibility of extracting all the cells represented in the images
- robustness due to the ANN application which allows resolving ambiguity of closed or touching cells.

An example of the last characteristic is shown in Fig. 9, where it can be seen how two cells that are clustered as a unique region by the FCM are well separated by the ANN thanks to the individuation of cytoplasm.

Different cases in which the ambiguity due to cell closeness is solved are shown in Fig. 10.

Due to the staining technique, two or more cells can touch each other along the cytoplasm. This ambiguity is easily solved by our classification method. In rare cases, two nuclei can overlap in great portions of their boundaries and also their appearance, after staining, can be very similar. In these cases (see Fig. 11), our algorithm is not able to overcome completely the problem (also human experts are not able to distinguish clearly the cells). In such critical situations, we could face the problem considering the derivability properties of the cells contours curves.

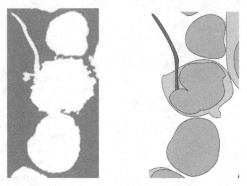

Fig. 9. Example showing the robustness of the proposed method: (*left*) rough segmentation obtained by FCM that individuates a unique region corresponding to three different cells; (*right*) result of the ANN algorithm where the cells are correctly separated by classifying pixels in cell body, cytoplasm and artifact (see Fig. 8 for explanation of colors). Example of the robustness of the proposed method: the image on the right shows the classification of the connected region of the image on the left by the ANN algorithm.

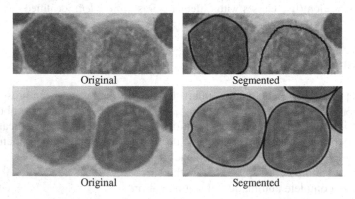

Original   Segmented

Original   Segmented

Fig. 10. Examples in which the ambiguity due to cell closeness is solved by the neural classification

In fact, we can identify the two singularity points P_1, P_2 (see Fig. 12) that are located in the two opposite semi-planes individuated by the maximal inertial axis of the segmented global structure, and, then, apply a curve prediction algorithm, determining independently the possible contours of each cell. Then, the actual contours of the two merged cells could be computed by means of a deformation process of the original boundaries in the region fixed by the two predicted curves. This separation procedure is under development, even if the very limited statistical occurrence of overlapping cases has not a relevant role in our problem.

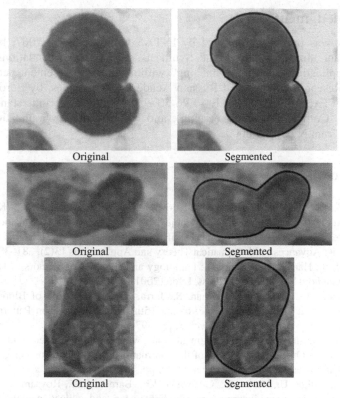

Fig. 11. Examples of cases in which the overlapping ambiguity is not solved by the classification procedure. In the bottom case, it is difficult to establish if the segmented structure is a single cell or two overlapping cells. Three cases of overlapping cells. In the bottom case, it is really difficult to establish if the segmented structure is a single cell or two overlapping cells.

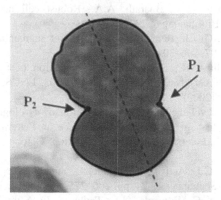

Fig. 12. Example of information that can be used in the separation procedure of two overlapping cells: the dashed line is the maximum inertial axis of the segmented global structure, the two points P_1 and P_2 are the two singularity points individuated along the contour

Acknowledgment

This work was partially supported by INTAS Grant N 04-77-7067, by the Cooperative grant "Image Analysis and Synthesis: Theoretical Foundations and Prototypical Applications in Medial Imaging" within bilateral agreement between Italian National Research Council and Russian Academy of Sciences, by European Project Network of Excellence MUSCLE – FP6-507752 (Multimedia Understanding through Semantics, Computation and Learning), and by the Russian Foundation for Basic Research Grant N 05-07-08000.

References

1. Churakova, Z.V., Gurevich, I.B., Jernova, I.A., Kharazishvili, D.V., Khilkov, A.V., Nefyodov, A.V., Sheval, E.V., Vorobjev, I.A.: Selection of Diagnostically Valuable Features for Morphological Analysis of Blood Cells. Pattern Recognition and Image Analysis: Advances in Mathematical Theory and Applications 13(2), 381–383 (2003)
2. Jaffe, E.S., Harris, N.L., Stein, H.: Pathology and Genetics of Tumors of Haematopoietic and Lymphoid Tissues. IARC Press, Lyon (2001)
3. Di Rubeto, C., Dempster, A., Khan, S., Jarra, B.: Segmentation of Blood Image using Morphological Operators. In: Proc. of the 15th Int. Conference on Pattern Recognition, Barcelona, Spain, September 3-8, vol. 3, pp. 397–400 (2000)
4. Anoraganingrum, D.: Cell Segmentation with Median Filter and Mathematical Morphology Operation. In: Proc. of the International Conference on Image Analysis and Processing, Venice, Italy, September 27-29, pp. 1043–1046 (1999)
5. Lin, G., Adiga, U., Olson, K., Guzowski, J.F., Barnes, C.A., Roysam, B.: A hybrid 3D watershed algorithm incorporating gradient cues and object models for automatic segmentation of nuclei in confocal image stacks. Cytometry A. 56(1), 23–36 (2003)
6. Adiga, U.P.S., Chaudhuri, B.B.: An efficient method based on watershed and rulebased merging for segmentation of 3-D histopathological images. Pattern Recognition 34(7), 1449–1458 (2001)
7. Mouroutis, T., Roberts, S.J., Bharath, A.A.: Robust cell nuclei segmentation using statistical modeling. BioImaging 6, 79–91 (1998)
8. Wu, H.S., Barba, J., Gil, J.: Iterative thresholding for segmentation of cells from noisy images. J. Microsc. 197, 296–304 (2000)
9. Karlosson, A., Strahlen, K., Heyden, A.: Segmentation of histological section using snakes. In: Bigun, J., Gustavsson, T. (eds.) SCIA 2003. LNCS, vol. 2749, pp. 595–602. Springer, Heidelberg (2003)
10. Murashov, D.: Two-level method for segmentation of cytological images using active contour model. In: Proc. of the 7th Int. Conference on Pattern Recognition and Image Analysis, PRIA-7, St. Petersburg, Russian Federation, October 18-23, vol. III, pp. 814–817 (2004)
11. Cootes, T.F., Taylor, C.J.: Active shape models – 'smart snakes'. In: Proc. of the British Mach. Vision Conf., September 21-24, pp. 266–275. Springer, Leeds, UK (1992)
12. Cootes, T.F., Beeston, C., Edwards, G.J., Taylor, C.J.: A unified framework for atlas matching using active appearance models. In: Kuba, A., Samal, M. (eds.) Proc. of Information Processing in Medical Imaging. LNCS, pp. 322–333. Springer, Germany (1999)

13. Ghebreab, S., Smeulders, A.W.M.: Strings: Variational Deformable Models of Multivariate Continuous Boundary Features. IEEE Transactions on Pattern Analysis And Machine Intelligence 25(11), 1399–1410 (2003)
14. Perner, P., Perner, H., Jänichen, S.: Recognition of Airborne Fungi Spores in Digital Microscopic Images. Artificial Intelligence in Medicine 36(2), 137–157 (2006)
15. Bezdek, L.C.: Pattern Recognition with Fuzzy Objective Function Algorithm. Plenum Press, New York (1981)
16. Rumelhart, D.E., Hinton, G.E., Williams, R.J.: Learning internal representations by error propagation. Parallel Distributed Processing, pp. 318–362. MIT Press, Cambridge, MA (1986)
17. Riedmiller, M., Braun, H.: A direct adaptive method for faster backpropagation learning: the RPROP algorithm. In: Proc. of the IEEE International Conference on Neural Networks – ICNN, San Francisco, CA, USA, March 3 – April 1, pp. 586–591 (1993)

Real-Time Measurement and Analysis of Translational and Rotational Speeds of Moving Objects in Microscope Fields

Primo Coltelli[1], Mauro Evangelisti[2], Valtere Evangelista[2],
and Paolo Gualtieri[2,3]

[1] CNR ISTI, via Moruzzi 1, 56124 Pisa Italy
[2] CNR Istituto di Biofisica, via Moruzzi 1, 56124, Italy
[3] to whom correspondence should be addressed

Abstract. This paper describes a digital system designed for the automatic detection and measurement of the velocity of moving objects in images acquired by means of a common TV-camera mounted onto a microscope. The main characteristics of this system are the following: 1) it can perform a real-time gray level difference between two successive frames in order to detect moving objects and to suppress stationary objects (subtraction procedure); usually the delay between two successive frames varies linearly from 40 msec to 1920 msec; 2) it reduces the size of images resulting from the subtraction procedure (difference images) and stores them in the frame memory; the result of these operation, all performed in real-time, is a film of time sequences; 3) it performs an automatic labelization in order to recognize the moving microorganisms and to calculate their area in each difference image; 4) it calculates and plots the variation of the average area of the cells moving in the microscope field; 5) it completes the analysis in few seconds.

1 Introduction

The exact determination of the speed parameters of swimming microorganisms can be a very useful tool for the study of both behavioral and physiological aspects of motility, that is an essential. Speed parameters can be obtained by means of photomicrographic [1] and cinematographic techniques [2]. These methods, however, turn out to be time consuming. Statistical counting techniques can be utilized as well [3], but they are tedious, and prone to human errors. Other methods utilize more sophisticated techniques such as spectral analysis of the light scattered by the microorganisms [4], or analogic elaboration of the video signal of a TV-camera mounted onto a microscope [5]. In these cases, however, the instrumentation necessary for speed parameter determination has the drawback to be not portable. A further alternative is represented either by the digital tracking microscope, which can determine speed parameters by reconstructing the entire movements of swimming microorganisms [6], [7], [8] or by the simple method we will describe. This method automatically and in real time determines the speed of swimming microorganisms. The biflagellate algae *Dunaliella salina* has been used as experimental subject.

P. Perner and O. Salvetti (Eds.): MDA 2006/2007, LNAI 4826, pp. 128–135, 2007.
© Springer-Verlag Berlin Heidelberg 2007

Substantially, this method utilizes the subtraction operation, which has already been used by other authors for the detection of motion [9]; however our procedure performs automatically and in real-time both the detection of the moving microorganisms and the determination of their speed parameters. Our results are consistent with previous published speed data of *Dunaliella salina* obtained with the other methods [10].

2 Materials and Methods

A Pulnix TM860 (Pulnix, USA) CCD video camera was mounted onto a Zeiss Axioplan microscope (Zeiss, Germany) equipped with 16x and 60x objectives and 100W halogen lamp as light source. Cells were placed in a small chamber obtained by fixing a PVC ring onto a microscope slide. The chamber was closed by means of a cover slip so as to avoid sample drying-out. The microorganisms can freely swim within a narrow layer of growth medium placed between a slide and a cover slip.

The signal of the camera was the input of a FG100 AT Frame Grabber (Image Technology, USA) plugged into a Pentium V personal computer 750MHz clock. For the translational speed determination experiment, a sequence of images taken at known intervals of time was acquired, stored, and processed using the automatic procedure of Gualtieri and Coltelli [11]. For the rotational speed experiment, the light reflected by the cell eyespot was measured. The experimental set-up was the same used previously, with the addition of a custom-made slide. This custom-made slide allowed the lateral illumination of the cell sample by means of an optical fiber delivering the light coming from a Schott KL1500 fiber optic illuminator (Schott, Germany).

Photographs were recorded with an Olympus Camedia C-30303 digital camera (Olympus, Japan) mounted on the Zeiss Axioplan microscope (Zeiss, Germany).

3 Operation Procedures

Real time detection of microorganisms under the microscope is performed by differencing continuously each frame of the video image from a previous frame, with a variable delay, during the acquisition process. This operation is made possible by programming the 12-bit input Look-up-table (LUT) of the board. This LUT, which is located between the digitization circuit and the frame memory, transforms the image before it is stored into the frame memory. Thanks to a feed-back circuit between the frame memory and the LUT, operations are made on combinations of stored and newly-acquired data. We program the LUT in order to move the six most significant bits from the A/D converter (the newly acquired data) to the six most significant bits in the frame memory; then the LUT subtracts the same six most significant bits of A/D data from those previously stored in the six most significant bits of the frame memory (the previous frame); the resulting six bits are then stored in the least significant bits of the frame memory. The resulting difference image is always available in the lower six 1-bit planes of the frame memory; while the upper six 1-bit planes contain the most recent data of the A/D converter, which are used as input for the next frame subtraction. In the case that the images of a moving cell in two successive frames are partly overlapping, the subtraction operation gives a zero value for the overlapping region of the cell and for the background, a negative value for that part of the cell image which is present only in the previously acquired frame and a positive value for that part of the cell image which is present only in the newly acquired frame. In order to follow the increasing of the cell image, which will increase up to the whole cell size during the delay progression, we program the LUT to clip to zero the negative value, (Figure 1).

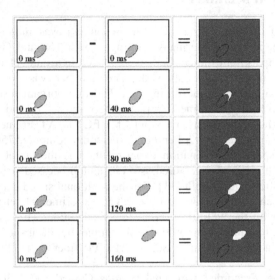

Fig. 1. Subtraction operation

In order to store several difference images in real time in the frame memory, we could reduce the spatial resolution of the image being acquired by means of the hardware Zoom. In order to store the (reduced) image into its proper position of the frame memory, the X and Y coordinates of its origin are shifted in real time by means of Pan and Scroll operations. In this way we can store images, by moving the coordinates of their origin toward the right and downward. At the end of this procedure, the frame memory is displayed as a patchwork of (reduced) images. In order to identify moving cells and to extract its features such as baricenter coordinates, contours, axis and areas, a segmentation and labelization procedure is applied to each difference image [11]. This procedure lasts 2 seconds for the whole memory.

4   Results and Discussion

Figures 2a and 2b represents 10-images time sequence (400 msec). Each difference image is represented as a framed image. The number visible in the first column represents the delay between the two frames on which the system has performed the difference. The system acquires a couple of frames utilizing for each frame six bits of the frame memory and performs the subtraction operation as previously explained. Usually we choose a delay that varies linearly, but the delay can progress in a different way as well. In our case, because of the CCIR standard, the delay between two successive frames is 40 msec, or a multiple of 40 msec. The first image (40 msec) represents the real time difference between the first acquired frame and the second acquired frame; the second image (80 msec) is the real time difference between the third acquired frame and the fifth acquired frame. The position of the frames in the successive couples can be easily extrapolated by the delay number. After subtraction every difference image is placed in its proper position of the frame memory in real time by means of pan and scroll operations. For the determination of the translational speed value of the cell we have to measure the distance covered by the cell and the time lapse; if no stimuli are applied to the environment, the swimming speed of the cells can be considered constant. Therefore, the time a cell takes to cover a distance equal to its long axis can be used for the determination of its speed. As the difference procedure presented in Figure 1 suggests, until the cell doesn't cover a distance equal to its size, the area value will be lower than the real one. Greater the delay between the frames, less two successive images of the cell are superimposed; there is a delay for which the subtraction operation gives two separate images of the same cell, i.e. the area value of the cell is the real value. A higher delay between the two frames still separates the two images of the cell, but the area value of the cell will remain constant. In Figure 2a and 2b, the cells, which are represented by the whitish areas, can be hardly recognized in the first frames because the difference between two successive images of the same cell consists of a small agglomerate of pixels. In the last frame of the same figure, the cells are instead easily recognizable, because in this case the difference between the two images of the same cell is the whole cell area. In order to determine quantitatively the cell area variation every reduced image is segmented and labelized, (last column of Figure 2a, 2b). Because of the reduced thickness of the medium, the swimming path of the cells is planar, i.e. the cells are always focused. The average area of the cells moving in the field is calculated and the detected cells are contoured. For each reduced image a pre-established standard deviation value determines the selection of an area value range. Therefore, touching cells are automatically rejected because their area is too big; similarly, small agglomerate of pixels, produced by the subtraction operation in the case of moving cells which intrudes onto the area formerly occupied by a different cell, or when a cell enters the field of view between the two frame on which the subtraction operation is performed, are rejected. The variation of cell number in the microscope field during the acquisition is not critical for the analysis, because we calculate the average area of the labeled cells present in each difference images. Twenty sequences are measured and the calculated areas averaged.

Figure 3 shows the plot of the average cells area versus the delay progression. By interpolating the data of this plot, we obtain two intersecting straight lines. The first line shows that the average area value increases with the increasing of the delay because the overlapping of two successive images of the same cell decreases. The second lines shows that the average area value becomes steady because there is no more overlapping between the two successive images of the same cell. The intersection of these two lines identifies the time delay which has to be used for the determination of the exact swimming speed of the microorganisms. In our case about

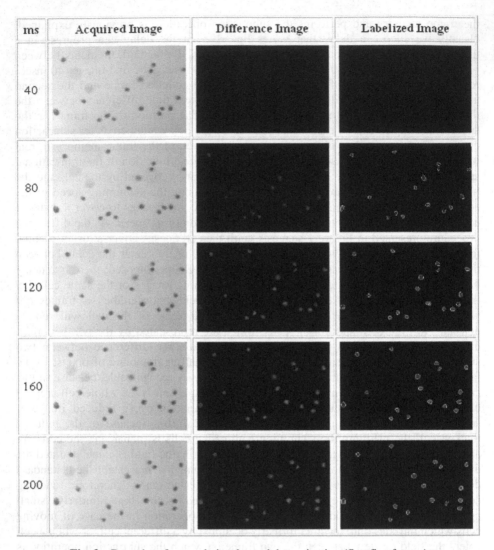

ms	Acquired Image	Difference Image	Labelized Image
40			
80			
120			
160			
200			

Fig. 2a. Procedure for translational speed determination (first five frames)

150 msec is the time *Dunaliella* cells need to cover a distance equal to its long axis. A quantitative determination of the long axis of *Dunaliella*, by means of our labelization procedure, gives an average value of about15 μ. Previous studies reported a velocity of 100 μ/sec for *Dunaliella* cells, [12], therefore we can state that our system gives a correct evaluation of the swimming speed of this microorganism.

To investigate the rotational speed we store in the computer memory the frames acquired under lateral illumination as described in the Material and Methods section. The eyespot of Chlorophyta such as *Dunaliella* is a quarter-wavelength multi layered organization of osmophilic granules, which reflects very efficiently the light that strikes upon it. As the cell moves, we can detect this brilliant spot and verify if the cell rotates or not. For the wild type of *Dunaliella* frames were acquired every 40 ms, were thresholded and labelized so the eyespot is recognized as present in the image, (Figure 4).

The resulting duty cycle from a 600 ms recording shows that these cells rotate with a frequency of 8 Hz. (Figure 5).

The time resolution of our system, which is 40 msec, can be considered sufficient to determine speed parameters of moving microorganism, as the study of physiological aspects of motility is usually based on the microscope observation of these phenomena. Due to its integration time, the human visual system has a time resolution of 250 msec, which is 6-time greater than that of our system [13]. Because the problem to be solved is the quantitative determination of visual phenomena, our system can be considered quite adequate for this purpose.

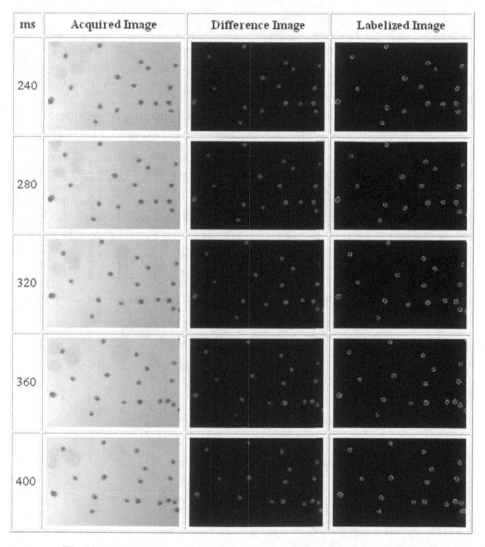

Fig. 2b. Procedure for translational speed determination (last five frames)

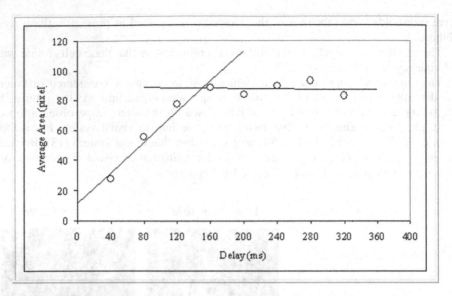

Fig. 3. The plot of the average area value of the cells vs. the delay between frames

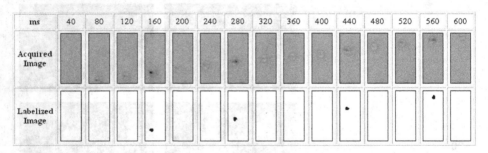

Fig. 4. Procedure for rotational speed determination

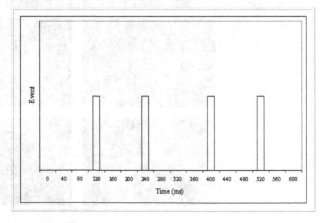

Fig. 5. The plot of the event of the eyespot detection vs. the elapsed time 600

References

1. Gibbons, B.H., Gibbons, I.R.: Flagellar movement and adenosin triphosphatase activity in sea urchin sperm extracted with triton X-100. J. Cell. Biol. 54, 75–82 (1972)
2. Phillips, D.M.: Comparative analysis of mammalian sperm motility. J. Cell. Biol. 53, 561–565 (1972)
3. Ojakian, G.K., Katz, D.F.: A simple technique for the measurement of swimming speed of Chlamydomanas. Exptl. Cell. Res. 81, 487–491 (1973)
4. Ascoli, C., Barbi, M., Frediani, C., Mure, A.: Measurements of Euglena gracilis motion parameters by laser light scattering. Biophys. J. 24, 585–599 (1978)
5. Biedert, S.W., Barry, W.H., Smith, T.W.: Inotropic effects and changes in sodium and calcium contents associated with inhibition of monovalent cations active transport by ouabain in cultured myocardial cells. J. Gen. Physiol. 74, 479–494 (1979)
6. Yachida, M., Asada, M., Tsuji, S.: Automatic analysis of moving images. IEEE Patter. Anal. Match. Intel. 3, 12–20 (1981)
7. Kondo, T., Kubota, M., Aono, Y., Watanabe, M.: A computerized video system to automatically analyze movements of individual cells and its application to the study of circadian rhythms in phototaxis and motility in Chamidomanas reinhardtii. Protoplasma Suppl. 1, 185–192 (1988)
8. Evangelista, V., Barsanti, L., Passarelli, V., Gualtieri, P. (eds.): From Cell to Proteins: Imaging Nature Across Dimensions. Springer, Dordlecht (2005)
9. Allen, R.: New direction and refinements in video-enhanced microscopy applied to problems in cell motility. In: Cowden, R.R., Harrison, F.W. (eds.) Advances in Microscopy, vol. 139, pp. 3–11. Alan Riss, New York (1983)
10. Vismara, R., Verni, F., Barsanti, L., Evangelista, V., Gualtieri, P.: A short flagella mutant of Dunaliella salina (Chlorophyta, Chlorophyceae). Micron 35, 337–344 (2004)
11. Coltelli, P., Gualtieri, P.: A procedure for the extraction of object features in microscope images. J. Biomed. Comput. 25, 169–176 (1990)
12. Barsanti, L., Gualtieri, P.: Algae, Anatomy, Biochemistry, and Biotechnology. Taylor and Francis, Boca Raton (2006)
13. Rose, A.: Vision: Human and Electronic. Plenum Press, New York (1977)

A General Approach to Shape Characterization for Biomedical Problems

Davide Moroni[1], Petra Perner[2], and Ovidio Salvetti[1]

[1] Istituto di Scienza e Tecnologie dell'informazione, ISTI-CNR, Pisa, Italy
davide.moroni@isti.cnr.it, ovidio.salvetti@isti.cnr.it
[2] Institute of Computer Vision and Applied Computer Sciences, Leipzig, Germany
pperner@ibai-institut.de

Abstract. In this paper, we present a general approach to shape characterization and deformation analysis of 2D/3D deformable visual objects. In particular, we define a reference dynamic model, encoding morphological and functional properties of an objects class, capable to analyze different scenarios in heart left ventricle analysis.

The proposed approach is suitable for generalization to the analysis of periodically deforming anatomical structures, where it could provide useful support in medical diagnosis. Preliminary results in heart left ventricle analysis are discussed.

1 Introduction

Deformable structures arise frequently in human anatomy and, in many cases, their deformation modes are of key importance in understanding the functional properties of the related organs and assessing their health-state. The main example is given by cardiac dynamic analysis, since many heart pathologies are correlated to the deformation pattern of the organ. In cardiac analysis, well-established imaging techniques are of great support in medical diagnosis, since they allow to acquire video sequences of the heart, from which its dynamical behavior can be inferred. However, the interpretation of the acquired data (temporal sequences of 2D/3D images, possibly from different imaging modalities) is difficult or, at least, time consuming; in daily practice, sometimes, physicians extract the most salient frames from the video sequence (end diastole and systole) and perform direct comparison among images in the selected subset. It is likely that, considering the full video sequence, more precise and rich information about the state of the heart can be discovered.

Motivated by these problems and extending the works [1,2], we believe that it is fruitful to define, in some generality, the concept of periodically deforming visual objects (see section 2 for a precise definition) and to propose a methodological approach to their study.

Besides providing modules for structures reconstruction and characterization, that have their own importance in biomedical applications as automatic tools to speed up diagnosis, the main idea is to define a reference dynamic model of an objects class: this model can be understood as an encoding of morphological and

P. Perner and O. Salvetti (Eds.): MDA 2006/2007, LNAI 4826, pp. 136–145, 2007.

functional properties of a periodically deforming object during its full deformation cycle. In particular, shape changes and evolution of local object properties
are depicted in a coincise form in the reference dynamic model, thus allowing
for deformation analysis and deformation pattern classification.

The paper is organized as follows. In section 2 we define the class of objects we
are interested in, making explicit the necessary assumptions. Then in section 3,
the proposed approach is outlined and its basic modules leading to the reference
dynamic model are described in detail. More precisely three modules are considered: *object reconstruction* (sec. 3.1), in which every object is reconstructed in
Euclidean space as a collection of manifolds, *object characterization* (sec. 3.2),
in which local shape descriptors and functional features are coded into property
functions and, finally, *deformation pattern assessment* (sec. 3.3) where the reference dynamic model is actually built. Preliminary results in heart dynamic
analysis are then presented in section 4, whereas conclusions and directions for
further work are briefly discussed in section 5.

2   Periodically Deforming Visual Object

A visual object O embedded in the background space $\Omega \subset \mathbb{R}^d$ $(d = 2, 3)$ is a
collection

$$O = \{(V^\alpha, P^\alpha)\}_{\alpha=1,2,\ldots,k}$$

where each V^α is a smooth manifold (possibly with boundary) embedded in Ω
and $P^\alpha : V^\alpha \to \mathbb{R}^{d(\alpha)}$ is a smooth *properties function* assuming its values in a
suitable properties space.

The smoothness assumption is a quite common hypothesis in computational
anatomy (see e.g. [3]) and it is satisfied in practice to a large extent; it implies for
example that differential geometric properties (like normals, curvatures,...) can
be computed everywhere. We use, moreover, collection of manifolds -instead of
a single one- to be able to describe object subparts (possibly of different dimensionality) by attaching them specific salient attributes via a dedicated properties
function. For example, in heart left ventricle modelling, the object of interest is
the myocardium, that can be modelled as a 3D manifold, whose boundaries are
two surfaces: the epicardium and the endocardium. It is convenient to attach to
the boundary surfaces a different (actually richer) set of attributes than those
used for internal points.

A deforming visual object $\mathcal{O} = (O_t)_{t=1,2,\ldots}$ is a temporal sequence of visual
objects satisfying some smoothness constraint. Each $O_t = \{(V^\alpha, P^\alpha)\}_{1\le\alpha\le k}$
should be regarded as the *snapshot* of the deforming object at time t.

We require that each manifold V_t^α appearing in the snapshot at time t can
be smoothly deformed into V_{t+1}^α in the subsequent snapshot. Tears or crack of
any object subpart are, therefore, ruled out; moreover, in such a way, we avoid
dealing with changes in topology, that would require to model shape transitions,
a task far beyond our present scopes.

Finally, a periodically deforming visual object is a deforming object for which
there exists an integer T such that $\forall t : O_t = O_{t+T}$. In other words, the deforming

object depicts a periodic motion; thus, a periodically deforming object is characterized by a finite list of snapshots $(O_1, O_2, \ldots, O_T)$, which will be referred to as its deformation cycle.

We make a final assumption about the data available to describe a periodically deforming visual object. It is assumed that a sufficiently rich set of synchronous signals and images, possibly from different modalities, has been acquired so as to represent faithfully a physical body or phenomenon of interest. In particular, the data set should include at least one 2D/3D image sequence $(I_t)_{1 \leq t \leq T}$, from which morphology and regional properties of the object can be inferred.

3   Methodology Definition

With the previous assumptions, a reference dynamic model of an object of interest is constructed by coding the dynamics of the object in a rich representation of its shape and functional properties.

The approach consists in three modules, each one performing specific tasks. Essentially, the first two modules are dedicated to extract a suitable periodically deforming visual object from image data. Then the periodically deforming visual object is analyzed and used to construct the reference dynamic model. A more precise outline of the modules used to obtain the aforementioned model is as follows:

Object reconstruction: For each phase t, the collection of manifolds $\{V_t^\alpha\}$ is identified and reconstructed in 2D/3D space by applying neural algorithms to the image sequence $(I_t)_{1 \leq t \leq T}$;

Object characterization: Morphological features and dynamic descriptors are extracted and coded in a property function P_t^α that for each point x of the manifold V_t^α returns the property vector $(P_1^\alpha(x), \ldots, P_m^\alpha(x))$, where each P_i^α represents one of the selected features;

Deformation pattern assessment: Suitable and significant shape descriptors are extracted and spatial distribution of the property functions are evaluated in order to obtain a description of the object dynamics.

In the following sections, these steps are described in more details.

3.1   Object Reconstruction

The 3D reconstruction of the visual object $\mathcal{O}$ is achieved via voxelwise classification, that is by labeling each voxel in the image domain with semantic classes which describe voxel membership to the collection of manifolds $\{V_t^\alpha\}$.

The classification is performed applying an advanced neural architecture to a set of extracted features. The involved features can be divided into two classes. First, low-level features are considered: they are context-independent and do not require any knowledge and/or pre-processing. Some examples are voxel position, gray level value, gradients and other differentials, texture, and so forth. Middle-level features are also selected, since voxel classification can benefit from more accurate clues, specific of the problem at hand. In particular, if an intrinsic reference system can be individuated to describe the object shape, it can be used to define a relative voxel position. If, in addition, a priori information about the object shape is available, a reliable clue for detecting edges in the images is given by the gradient along the normal direction to the expected edge orientation.

Moreover, a multiscale approach is adopted: the features are computed on blurred images, supplying information about the behavior of the voxel neighborhood, which results in a more robust classification.

The set of selected features are processed to accomplish the voxel classification by means of a Multilevel Artificial Neural Network (MANN), which assures various computational advantages [4]. For each voxel x, its computed features vector is splitted into vectors $\mathcal{F}_k(x)$, each one containing features of the same typology and/or correlated. Then each $\mathcal{F}_k(x)$ is processed by a dedicated classifier based on an unsupervised Self Organizing Maps (SOM) architecture. The set of parallel SOM modules constitutes the first level of the MANN which aims at clustering each portion of the feature vector into crisp classes, thus reducing the computational complexity. The output of this first level is then passed to a second and final level, consisting in a single Error Back-Propagation (EBP) module, which supplies voxel classification.

Its output describes voxel membership to the various manifolds V_t^α in the collection $\{V_t^\alpha\}_{1 \leq \alpha \leq k}$.

3.2 Object Characterization

The reconstructed object is further characterized by assigning a significant properties function $P_t^\alpha : V_t^\alpha \to \mathbb{R}^{d(\alpha)}$ to each manifold V_t^α.

Three types of properties are considered:

- intensity based properties;
- local shape descriptors;
- local dynamic behavior descriptors.

Examples of properties of the first type are gray level value, gradients, textures and so on. They are extracted form the image sequence I_t –the one which leads us to object reconstruction. If data collected from other imaging modalities are available, after performing registration, we can fuse this information to further annotate the object (for example, in the case of the heart, information regarding perfusion and metabolism, obtained e.g. by means of PET imaging, can be referred to the reconstructed myocardium). Geometric based properties, belonging to the second type, are extracted directly from the collection of manifolds $\{V_t^\alpha\}$, and are essential to describe locally the shape of the object. Again, we may distinguish between context independent features (automatically computable for every manifold of a given dimensionality, such as Gaussian and mean curvature for surfaces) and problem-specific properties.

Finally, the local dynamic behavior may be described by properties borrowed from continuous mechanics (such as velocity field and strain tensor); they, however, require, at least, local motion estimation, that we haven't pursued yet.

3.3 Deformation Pattern Assessment

The periodically deforming visual object obtained in the previous steps can be used to assess the dynamic behavior of the object and identify its deformation pattern. However, the voxelwise characterization of the reconstructed objects is not suited for state assessment. Indeed, the given description of the whole objects (collection of manifolds described by functions) has a dimensionality far too high to make the problem computationally feasible. Moreover, it would be essential to be able to compare anatomical structures belonging to different patients and, at the moment, the idea is to use a deformable model (given for example by mass-spring models [5]) and to normalize every instance of anatomical structure to that model: in this way anatomical structures (belonging to the same family) are uniformly described and can be then compared.

Combining these two issues, we should look for a new set of 'more intrinsic' features $\mathcal{F}_t$ that should be enough simple and, at the same time, capturing essential information about the objects.

To obtain these new kinds of features, global information about the objects can be extracted from the properties function, without introducing any model. For example, one may consider the 'property spectrum', by which we mean the probability density functions (PDF) of a given component of the property function $P_t^\alpha(\cdot)$. This consists in a function capturing how the property is globally distributed; thus, comparison of different property spectra is directly feasible; to reduce dimensionality, moreover, it is effective to compute the momenta of the PDF (mean, variance,...).

However, properties spectrum does not convey any information at all about regional distribution of the property. In practical situation, this is a drawback which cannot be ignored: for example, a small 'highly abnormal' region may not affect appreciably the PDF, but its clinical relevance is, usually, not negligible. Hence, spatial distribution of properties has to be analyzed; in some cases, approaches which do not need a refined model of the object (e.g., Gaussian image, spherical harmonics or Gabor spherical wavelets) may be suitable. However, in general one should define a model of the objects (whose primitives -elementary bricks- are regions, patches or landmarks) and then propagate it to the set of instances to be analyzed by using matching techniques. Then, we may consider the average of a property on regions or patches (or the value in a landmark) as a good feature, since comparisons between averages on homologous regions can be immediately performed.

Following this recipe, a vector of features $\mathcal{F}_t$ with the desired properties is obtained for each phase of the cycle. The deforming object is then described by the dynamics of the temporal sequence of feature vectors obtained at different phases of the deformation cycle.

A further fruitful feature transformation may be performed exploiting our assumptions on deformable visual objects. Indeed, the smoothness of deformations implies that a visual object has mainly low frequency excited deformation modes. We extend this slightly assuming that this holds true also for the features lists $(\mathcal{F}_t)_{1 \leq t \leq T}$. We assume that the fundamental frequency of the motion is also the main component of each feature tracked on time. With these assumptions, an obvious choice is given by the Fourier transform, followed by a low pass filter, which supplies a new features vector Θ.

The evaluation of the above mentioned parameters $\mathcal{F}_t$, at each phase t, implicitly codifies information regarding object dynamics. Actually, we avoid defining a complex model of the object kinematics and exploit its periodic characteristic by constructing a rich representation of each phase of the deformation cycle.

4   Results

An elective case study for the presented methodology is cardiac analysis, whose clinical relevance can be hardly overestimated. We restrict our analysis to the left ventricle (LV) that, pumping oxygenated blood around the body, is the part of the heart for which contraction abnormalities are more clinically significant.

The proposed methodology is, of course, not universal, in the sense that there are some intrinsic limitations that prevent it to be potentially applied in any scenario. Indeed, our analysis is limited to a single deformation cycle and so only pathologies that affect every deformation cycle can be considered. Moreover, we require that physiological and (selected) pathological states induce different feature dynamics. This requirement is not too restrictive; actually, it is well known that many pathologies are correlated to abnormal shape patterns at end systole.

The LV structure is modelled as a 3D manifold (the myocardium) with boundary. The boundary has two connected components which are the surfaces corresponding to epicardium and endocardium.

We describe henceforth how the steps of the methodology are applied. First, the deformable visual object structure is extracted from the available data, consisting in a sequence of short axis gradient echo MR images, acquired with the FIESTA, GENESIS SIGNA MRI device (GE medical system), 1.5 Tesla, TR = 4.9 ms, TE = 2.1 ms, flip angle 45° and resolution $(1.48 \times 1.48 \times 8)$ mm. Sets of $n = 30$ 3D scans, consisting of $k = 11$ 2D slices, were acquired at the rate of 30 ms for cardiac cycles [diastole-systole-diastole]. Various clinical cases were considered, for a total of 360 scans, corresponding to 12 cardiac cycles.

To perform reconstruction, we first used a pre-processing step devoted to the automatic localization of the left ventricle cavity (LVC) [6].

The located LVC is then exploited to define an Intrinsic Reference System (IRS), given by a hybrid spherical/cylindrical coordinates system. This choice is dictated by the fact that LV approximately resembles a bullet-shaped structure; moreover, in the IRS, image partial derivatives w.r.t. radial coordinate are an efficient clue for heart surfaces detection.

The IRS is used to extract the following features for voxel classification:

– Position w.r.t. IRS
– Intensity and Mean intensity (computed applying Gaussian filters)
– Gradient norm $\|\nabla I_t\|$
– Partial derivative in the radial direction $\frac{\partial I_t}{\partial r}$.

Using the 2-level ANN, voxels are classified on the basis of their features vector as belonging or not to epi- and endocardial surfaces. More in detail, the set of extracted features is divided into two vectors $\mathcal{F}_1$, $\mathcal{F}_2$ containing respectively position, intensity and mean intensity, and position, gradient norm and partial derivative $\frac{\partial I_t}{\partial r}$. The position w.r.t. IRS is replicated in both vectors because it reveals salient for clustering both features subsets. Then, the first level of the MANN consists of two SOM modules, which have been defined as 2D lattice of neurons and dimensioned experimentally, controlling the asymptotic behaviour of the number of excited neurons versus the non-excited ones, when increasing the number of total neurons [7].

A 8 × 8 lattice SOM was then trained, according to Kohonen's training algorithm[8], for clustering the features vector $\mathcal{F}_1$, while $\mathcal{F}_2$ was processed by a 10 × 10 lattice SOM.

A single EBP module has been trained to combine the results of the first level and supply the final response of the MANN. The output layer of this final module consists in two nodes, which are used separately for reconstructing the epicardium and the endocardium. Since each cardiac surface divides the space into two connected regions (one of which is bounded), each output node can be trained using the signed distance function with respect to the relative cardiac surface. In this way, points inside the surface are given negative values, whereas positive values are given to points in the outside. Henceforth the surface of interest correspond to the zero-level set of the output function.

Different architectures have been tested, finding the best performance for a network with only one hidden layer of 15 units, trained according to the Resilient Back-Propagation algorithm [9].

The voxel classification, supplied by the MANN, may be directly used for visualization purposes by using an isosurface extraction method, as shown in figure 1.

Characterization of the reconstructed structure is obtained annotating every voxel with intensity, Gaussian and mean curvature, wall thickness and IRS properties. In particular, Gaussian and mean curvature have been included as shape descriptors whereas wall thickness, which is a classical cardiac parameter, is one example of problem-specific property: it is defined as the thickness of the myocardium along a coordinate ray and it is expected to increase during contraction (since myocardium, being almost water, is, with good approximation, incompressible).

This characterization is translated in a more amenable form by computing properties spectrum and regional features. In computing spectrum, coordinates w.r.t. IRS have been disregarded, with the exception of radial coordinate; intensity has also been excluded. For any property only mean and variance have been considered. For computing regional features, so far, we used a popular model of the LV (see [10] for a review of 3D-cardiac modelling). In 2D, as shown in Figure 3, it is defined by the intersections of cardiac surfaces with a pencil of equally spaced rays. The 3D version is obtained by stacking the 2D construction along the axis of the LV.

Fig. 1. Different views of the rendered left ventricle at end diastole. The surfaces are obtained applying marching cubes on the two output functions of the network. To eliminate satellites, a standard island removing procedure is applied.

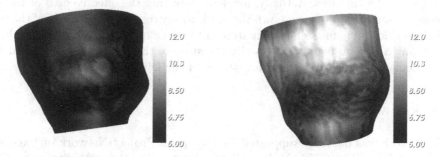

Fig. 2. Wall thickness at end diastole and systole, shown as an attribute of epicardial surface. Estimation is performed according to the centerline method and values are expressed in millimeters.

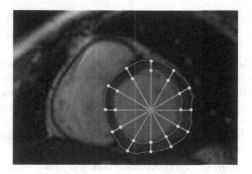

Fig. 3. The pencil of equally spaced rays used to computed local features

5   Conclusions and Further Work

In this paper, we define a reference dynamic model, encoding morphological and functional properties of an objects class, capable to analyze different scenarios in heart left ventricle analysis. In particular, a framework for the shape characterization and deformation analysis has been introduced for the study of periodically deforming objects.

This framework consists of several modules performing a) object reconstruction, b) object characterization, c) pattern deformation assessment. Solutions to specific tasks proposed in each module are, to a large extent, independent and may be combined with other methods, thus broadening the potential application field of the framework. In particular, an approach based on multi-level artificial neural network has been selected as a general purposes strategy for object reconstruction, motivated by the promising results presented in [4]. A quantitative evaluation of segmentation performance, based on comparison between images automatically segmented and images annotated by a committee of expert observers, however, is still in progress.

The elective case studies are represented by the analysis of heart deformable anatomical structures. Actually, for demonstrating the effectiveness of the proposed framework, we have shown the preliminary results in the study of the heart left ventricle dynamics. The next step will be to employ the obtained results for defining a general method to classify the state of the deformable object, and, in particular, the physio-pathological states of the left ventricle.

Acknowledgments

This work was partially supported by European Project Network of Excellence MUSCLE - FP6-507752 (Multimedia Understanding through Semantics, Computation and Learning) and by European Project HEARTFAID "A knowledge based platform of services for supporting medical-clinical management of the heart failure within the elderly population"(IST-2005-027107).

References

1. Colantonio, S., Moroni, D., Salvetti, O.: A methodological approach to the study of periodically deforming anatomical structures. In: AITTH 2005, pp. 32–36 (2005)
2. Colantonio, S., Moroni, D., Salvetti, O.: Shape comparison and deformation analysis in biomedical applications. In: Eurographics Italian Chapter Conference, pp. 37–43 (2006)
3. Grenander, U., Miller, M.I.: Computational anatomy: an emerging discipline. Q. Appl. Math. LVI(4), 617–694 (1998)
4. Di Bona, S., Niemann, H., Pieri, G., Salvetti, O.: Brain volumes characterisation using hierarchical neural networks. Artificial Intelligence in Medicine 28(3), 307–322 (2003)
5. Di Bona, S., Lutzemberger, L., Salvetti, O.: A simulation model for analyzing brain structures deformations. Physics in Medicine and Biology 48, 4001–4022 (2003)
6. Colantonio, S., Moroni, D., Salvetti, O.: MRI left ventricle segmentation and reconstruction for the study of the heart dynamics. In: IEEE ISSPIT, Athens, Greece, pp. 213–218. IEEE Computer Society Press, Los Alamitos (2005)
7. Di Bono, M., Pieri, G., Salvetti, O.: A tool for system monitoring based on artificial neural networks. WSEAS Transactions on Systems 3(2), 746–751 (2004)
8. Kohonen, T.: Self-Organizing Maps, 2nd edn. Springer Series in Information Sciences, vol. 30 (1997)

9. Riedmiller, M., Braun, H.: A direct adaptive method for faster backpropagation learning: The RPROP algorithm. In: Proc. of the IEEE Intl. Conf. on Neural Networks, pp. 586–591. IEEE Computer Society Press, San Francisco, CA (1993)
10. Frangi, A.F., Niessen, W.J., Viergever, M.A.: Three-dimensional modeling for functional analysis of cardiac images: A review. IEEE Trans. Med. Imaging 20(1), 2–5 (2001)

Statistical Analysis of Electrophoresis Time Series for Improving Basecalling in DNA Sequencing

Anna Tonazzini and Luigi Bedini*

Istituto di Scienza e Tecnologie dell'Informazione - CNR
Via G. Moruzzi, 1, I-56124 PISA, Italy
anna.tonazzini@isti.cnr.it

Abstract. In automated DNA sequencing, the final algorithmic phase, referred to as basecalling, consists of the translation of four time signals in the form of peak sequences (electropherogram) to the corresponding sequence of bases. Commercial basecallers detect the peaks based on heuristics, and are very efficient when the peaks are distinct and regular in spread, amplitude and spacing. Unfortunately, in the practice the signals are subject to several degradations, among which peak superposition and peak merging are the most frequent. In these cases the experiment must be repeated and human intervention is required. Recently, there have been attempts to provide methodological foundations to the problem and to use statistical models for solving it. In this paper, we exploit a priori information and Bayesian estimation to remove degradations and recover the signals in an impulsive form which makes basecalling straightforward.

1 Introduction

In automated sequencing, a reaction of extension from the initial primer of a given DNA strand generates a complete set of fragments in which the last base is marked with a fluorescent dye out of four different types, one for each type of base. Fragments are then sorted by length by means of electrophoresis and detected, as they pass under a laser, by four optical sensors, capturing the emission in the distinct wavelength ranges where the four dyes emit. The result is an electropherogram, that is four time series in the form of peak sequences, each representing the variation with time of the concentration of DNA fragments ending with the same base. Each peak in the four signals represents a base, its size is related to the number of DNA fragments of a given length, while its time location is related to the specific length and reflects the position of the base in the DNA strand under consideration. Basecalling is the final algorithmic phase

---

* This work has been supported by the European project Network of Excellence MUSCLE FP6-507752 (Multimedia Understanding through Semantics, Computation and Learning).

P. Perner and O. Salvetti (Eds.): MDA 2006/2007, LNAI 4826, pp. 146–155, 2007.

of automated sequencing, and consists in obtaining the DNA base sequence from the electropherogram by the ordered reading of the peaks.

The most popular commercial basecallers are the software developed by ABI [1], running on the ABI Prism sequencers, and Phred [5] [6], which has been used in the Genome Project. Both are based on peak detection algorithms refined with heuristics, and are very efficient when the peaks are well distinct and quite regular in spread, amplitude and spacing.

Unfortunately, in the practice data production is subject to several processes that lead to degradations of the electropherograms, particularly near the end of the sequence. Among those, the most important and frequent are peak superposition, known as cross-talk, due to the spectral overlapping between fluorescent dyes, and peak merging, known as diffusion, due to mobility shifts and deviations of the fragments in the gel. Signal leakage may also occur, resulting in secondary peaks. These degradations may seriously affect the performance of basecalling algorithms, and, in the current practice, they entail repeating the experiment, comparing the base sequence with that of the complementary strand [7], and manual editing.

The availability of economic DNA sequencers and reliable and fast basecalling algorithms, which allow to reduce as much as possible the intervention of human operators, is still an open issue and is especially important in order to cope with large scale sequencing of whole genomes, sequencing of the genomes of as many as possible species, comparative genomics and evolutionary studies, and the increasing diffusion of sequencing of individual DNA segments in the clinical practice. Furthermore, accuracy up to a single base would be essential for reliable locations of SNPs, and for the efficiency of gene prediction software, e.g. to avoid premature termination due to false stop codons.

In the literature, there have been several attempts to provide methodological foundations to basecalling, and to use statistical models which allow the incorporation of prior knowledge about the structure of the problem and the data directly into the basecalling algorithm, without resorting to heuristics [3], [8], [10], [11], [12]. In particular, in [8] hidden Markov models and Markov chain Monte Carlo methods are used.

In this paper, the problem of removing cross-talk and diffusion in electropherograms is formulated as one of joint blind source separation and blind deconvolution. In particular, Bayesian estimation and a priori information are exploited to recover the signals in an impulsive form which makes the task of basecalling straightforward.

2   Problem Formulation and Bayesian Estimation

In ideal conditions (i.e. same velocity for all fragments of a given length, fluorescence emission in four separated wavelength ranges) the ideal electrophoretic signal s_j, $j = 1, 2, 3, 4$, would be an impulse train, where the impulse locations identify the mutual positions of the bases of type j out of four types (namely, A, T, C and G) along the DNA strand under consideration, and the impulse magnitudes vary in time according to the changing color concentration, i.e the changing number of fragments of a given length. It is immediate to see that in such a case the task of basecalling would be straightforward. Conversely, the fragment mobility is a random process, subject to variations due to the nature of the experiment, and the $j-th$ measured signal x_j, $j = 1, 2, 3, 4$, represents the intensity of fluorescence emitted in one of four wavelength ranges, where the emission spectra overlap. The electropherograms can thus be considered as the result of the application to the ideal signals of two operators in cascade: a convolution with a kernel related to the mobility distribution (diffusion), followed by a mixture of the four signals, modelling the superposition of the fluorescence emission spectra in four wavelength ranges (cross-talk). The data model we consider is thus:

$$x_i(t) = \sum_{j=1}^{4} A_{ij} \left(h_j * s_j\right)(t) + b_i(t) \qquad \forall t, \quad i = 1, 2, 3, 4 \qquad (1)$$

where b_i is a noise term incorporating the left error sources, A is the 4×4 unknown cross-talk matrix, and h_j is the unknown impulse response which models the diffusion effect. This is related to the peak shape, depending on the casuality of fragment mobility. In general, h_j can be considered a Gaussian function, with unknown and time-varying variance. Indeed, longer fragments are more prone to mobility variations, so that it is expected that the variance of the impulse response slowly increases with time. Asymmetric, heavy-tailed peaks, due to deviations in the gel of long fragments, could be modelled as a mixture of Gaussians.

In the current practice, noise removal, cross-talk correction (also referred to as color separation) and deconvolution are performed off-line and separately, as steps of pre-processing of the electropherogram, prior applying basecaller software. As per the noise, this is assumed as constituted of two terms. A white, Gaussian term is suitable for modelling error sources such as fluorescent impurity in the gel, electronics and light scattering, and, since the actual DNA fluorescence is a very slowly varying signal, low pass filtering is usually employed to filter this noise out. Another noise term is the baseline, i.e. an error term due to a constant value of background fluorescence, which is modelled as a slowly increasing function of time. The correction of the baseline error is usually performed by removing a roughly constant waveform from the recorded signals.

To eliminate the cross-talk between the four channels of the electropherogram, usually a linear operator is applied. As already said, cross-talk is due to the overlapping of the outputs of the optical filters that separate the fluorescence from each of the four tags. This overlapping is linear and can be modelled through a mixing matrix. When this matrix is known, the cross-talk can be eliminated by applying its inverse to the data. The mixing matrix, however, is not known, and must be determined. To this purpose, techniques mostly based on analysis of the second order statistics of the signal have been proposed.

Given the data model of eq. 1, our aim is instead to jointly perform estimation of mixing and diffusion, and color separation and deconvolution, using a priori knowledge that one might have about the problem. Thus, recovering the ideal signals, i.e. removing the cross-talk and peak spreading effects, is seen as a problem of joint blind source separation and blind deconvolution. In a Bayesian framework, we propose a Maximum A Posteriori (MAP) estimate for the unknowns of the problem:

$$(\hat{\mathbf{s}}, \hat{A}, \hat{\mathbf{h}}) = arg \max_{\mathbf{s},A,\mathbf{h}} P(\mathbf{s}, A, \mathbf{h}|\mathbf{x}) = arg \max_{\mathbf{s},A,\mathbf{h}} P(\mathbf{x}|\mathbf{s}, A, \mathbf{h})P(\mathbf{s})P(A)P(\mathbf{h}) \quad (2)$$

where $P(\mathbf{x}|\mathbf{s}, A, \mathbf{h})$ is the noise distribution, and $P(\mathbf{s})$, $P(A)$, and $P(\mathbf{h})$ are the prior distributions for the three independent sets of variables. At present, we consider the noise term to be in its whole a white, Gaussian and stationary process. Although in our approach the baseline is considered incorporated in a generic Gaussian noise term, from experiments conducted on synthetic data we have seen that the proposed method is robust enough against non-stationary noise, whose variance is slowly increasing with time.

The prior adopted for the signals has been chosen on the basis of the minimum number of constraints one may reasonably enforce on the expected, ideal output of the electrophoresis process. In blind source separation, when as in our case both the mixing and the sources are unknown, a typical constraint which is enforced to sort out a solution from the infinite ones which fit the data, is statistical independence of the sources. This approach has given rise to a number of very efficient methods and algorithms known as independent component analysis (ICA) [2] [4] [9]. In our case, however, ICA is not suitable, since we know that the four electrophoretic signals should not be superimposed to each other. This means that the sources are actually dependent, but, at the same time, this information provides us with a very powerful constraint for efficiently bounding the problem. Thus, in our method, to obtain separation, at each time t only one signal out of the four is allowed to be non-zero. For deconvolution, we enforce positivity and minimum energy of the signals. Indeed, these constraints used together are able to produce super-resolution, and then are very effective for the deconvolution of impulsive signals. With respect to the estimation of the mixing and diffusion operators, we considered generic constraints for both A and $\mathbf{h}$. In particular, the adopted prior for A constrains its elements to be positive, while $\mathbf{h}$ is modelled as a Gaussian function and bounds on its variance are used.

The joint MAP estimation of eq. 2 is usually approached by means of alternating componentwise maximization with respect to the three sets of variables in turn:

$$\hat{\mathbf{h}} = arg \max_{\mathbf{h}} P(\mathbf{x}|\mathbf{s}, A, \mathbf{h})P(\mathbf{h}) \quad (3)$$

$$\hat{A} = arg \max_{A} P(\mathbf{x}|\mathbf{s}, A, \mathbf{h})P(A) \quad (4)$$

$$\hat{\mathbf{s}} = arg \max_{\mathbf{s}} P(\mathbf{x}|\mathbf{s}, A, \mathbf{h})P(\mathbf{s}). \quad (5)$$

where the priors $P(\mathbf{h})$, $P(A)$ and $P(\mathbf{s})$ are chosen in such a way to probabilistically enforce the over-mentioned constraints. We solve the above scheme via a Simulated Annealing algorithm in A and $\mathbf{h}$, alternated with deterministic updates for $\mathbf{s}$, based on gradient ascent.

3   Experimental Results

To quantitatively measure the performance of the proposed method, we carried out a number of experiments on synthetically generated DNA electropherograms. Two of such experiments are illustrated in Figures 1-3, for the noiseless and noisy cases, respectively. The data were generated by convolving four non superimposed impulse trains with Gaussian impulse responses, and then linearly mixing the four resulting signals. For each impulse train, the number of impulses, their locations and amplitudes, were chosen randomly, and the standard deviation of the corresponding impulse response was kept fixed along the sequence, in the assumption that diffusion can be considered stationary for short sequences.

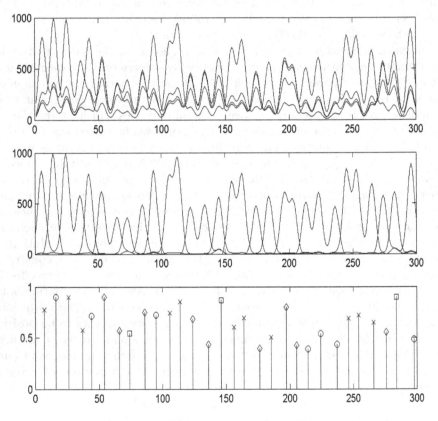

Fig. 1. Top: noiseless synthetic DNA sequencing data; Middle: color corrected data; Bottom: output from joint separation and deconvolution

Figure 1 shows the results of the method in the noiseless case. In particular, the top panel shows the very bad quality electropherogram considered as data, the middle panel shows the intermediate result of blind color separation, and the bottom panel shows the final signals reconstructed after both color separation

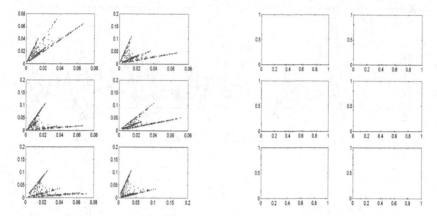

Fig. 2. Left: data scatterplot; Right: reconstruction scatterplot

and blind deconvolution. Note, however, that the algorithm directly produces the final reconstruction starting from the data alone, and the intermediate color separation result has been obtained by multiplying the data with the inverse of the estimated mixing matrix. In this case, the final reconstruction exactly reproduces the positions of the original impulse trains considered, apart from scale factors in the amplitudes. The symbols marking the different impulses indicate the four kinds of DNA bases (A, T, C, G). The original mixing matrix adopted for generating the data was:

$$A_{true} = \begin{bmatrix} 1.0000 & 0.4976 & 0.1277 & 0.2129 \\ 0.9536 & 1.0000 & 0.3723 & 0.2415 \\ 0.6725 & 0.7184 & 1.0000 & 0.3345 \\ 0.2725 & 0.2136 & 0.3266 & 1.0000 \end{bmatrix}$$

while the estimated one was:

$$A_{est} = \begin{bmatrix} 1.0000 & 0.4675 & 0.1348 & 0.2185 \\ 0.9525 & 1.0000 & 0.3697 & 0.2422 \\ 0.6862 & 0.6683 & 1.0000 & 0.3453 \\ 0.2717 & 0.2693 & 0.3014 & 1.0000 \end{bmatrix}$$

For comparison purposes, the two matrices has been rescaled by dividing each column for its highest value. The mean square error between A_{true} and A_{est} was 0.0217. The standard deviations of the four Gaussian impulse responses were estimated up to an accuracy of 0.001. Figure 2 shows the scatterplots of the data (left panel) and of the reconstructed signals (right panel). While a high degree of correlation is present between each couple of data signals, the reconstructions are perfectly uncorrelated.

In another experiment, shown in Figure 3, we added same noise to the convolved and mixed signals. This was a white, Gaussian process, with standard deviation slowly increasing with time, to simulate the baseline error. Also in

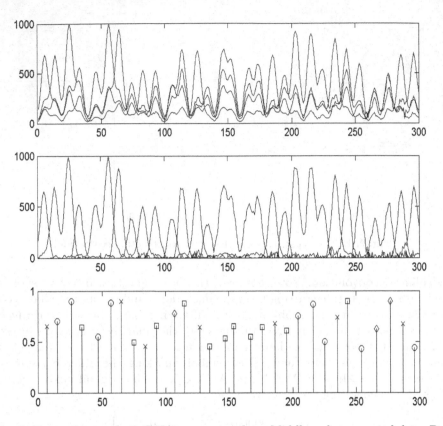

Fig. 3. Top: noisy synthetic DNA sequencing data; Middle: color corrected data; Bottom: output from joint separation and deconvolution

this case the reconstructions of the signals, the mixing matrix and the impulse response standard deviations were more than satisfactory, showing that the method is robust enough even against non-stationary noise.

Other experiments were conducted on real data, for which results from automatic sequencing machines were available. In particular, we performed tests on several segments of the genome of a *Gymnochlora* sp. alga. With our method, we obtained some improvements with respect to the performance of the commercial basecallers, even for high quality electropherograms. Figure 4 shows the result obtained on a segment for which the reliability of the calls of the commercial basecaller was very low. For this segment, the highly reliable sequencing of the complementary strand was available. We could thus perform a biological validation of the results, based on an estimate of the true sequence, obtained for complementarity from the dual strand. In particular, we observed that the sequence provided by the commercial sequencer contained seven errors (mainly missing bases, i.e. deletions), while ours only three errors. Finally, Figure 5 shows a short sequence of satisfactory quality where, however, the software running on the commercial automated sequencing machine produced an error in the interval 100-150 where the

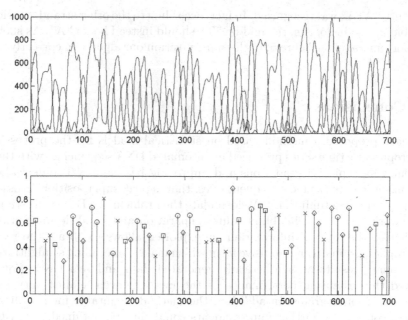

Fig. 4. Top: real DNA sequencing data; Bottom: output from joint separation and deconvolution

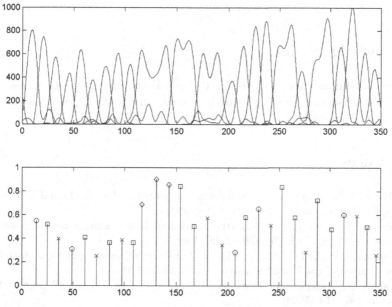

Fig. 5. Top: real DNA sequencing data; Bottom: output from joint separation and deconvolution

sequence "ATA" was recognized. In fact, according to the other strand, considered reliable by the biologists, the middle "T" should instead be an "A". As shown in the bottom panel of Figure 5, in the same position our algorithm correctly recognized an "A".

4   Conclusions

We have proposed a method based on statistical models for the processing of electrophoretic time series produced in automated DNA sequencing, with the aim at removing typical degradations and improving basecalling. The degradations we considered are the most frequent ones, that is peak superposition (cross-talk) and peak merging (diffusion). We formulate the problem in a Bayesian estimation framework as one of joint blind source separation and blind deconvolution. In particular, the a priori information we exploited allows to obtain the restored electropherograms in an impulsive form which makes basecalling straightforward. Preliminary results on synthetically generated data and real DNA sequences showed the promising performance of the method, even against very bad or noisy electropherograms. In addition, the method is suitable for handling any kind of constraint. Further improvements could thus be obtained, for instance, by including constraints on the number of allowed bases and bounds on the spacing between bases. Another possible extension could consists in enforcing the bases to be complementary with those of the other strand, when, as it often happens, this is available as well.

Acknowledgements

We are grateful to Paolo Gualtieri and Anna Frassanito of the Biophysics Institute, CNR, Pisa for providing us with the data and assisting us in the biological validation of the results.

References

1. ABI. ABI Prism.: DNA sequencing analysis software, User's Manual. Perkin Elmer Applied Biosystems, Foster City, CA (1996)
2. Amari, S., Cichocki, A.: Adaptive blind signal processing - neural network approaches. Proc. IEEE 86, 2026–2048 (1998)
3. Boufounos, P., El-Difrawy, S., Ehrlich, D.: Basecalling using hidden Markov models. Journal of the Franklin Institute 341, 23–36 (2004)
4. Comon, P.: Independent Component Analysis, a new concept? Signal Processing 36, 287–314 (1994)
5. Ewing, B., Hillier, L., Wendl, M., Green, P.: Base-calling of automated sequencer traces using Phred. I, Accuracy assessment. Genome Res. 8, 175–185 (1998)
6. Ewing, B., Green, P.: Base-calling of automated sequencer traces using Phred. II, Error probabilities. Genome Res. 8, 186–194 (1998)

7. Freschi, V., Bugliolo, A.: Computer-aided DNA base calling from forward and riverse electropherograms. In: Priami, C., Merelli, E., Gonzalez, P., Omicini, A. (eds.) Transactions on Computational Systems Biology III. LNCS (LNBI), vol. 3737, pp. 1–13. Springer, Heidelberg (2005)
8. Haan, N.M., Godsill, S.J.: Modelling electropherogram data for DNA sequencing using variable dimension MCMC. In: Proc. Int. Conf. on Acoustics Speech and Signal Processing - ICASSP, Instanbul, Turkey, pp. 3542–3545 (2000)
9. Hyvärinen, A.: Fast and Robust Fixed-Point Algorithms for Independent Component Analysis. IEEE Trans. Neural Networks 10, 626–634 (1999)
10. Li, L., Speed, T.P.: An estimate of the cross-talk matrix in four-dye fluorescence-based DNA sequencing. Electrophoresis 20, 1433–1442 (2000)
11. Li, L.: DNA sequencing and parametric deconvolution. Statistica Sinica 12, 179–202 (2001)
12. Pereira, M., Andrade, L., El-Difrawy, S., Manolakos, E.S.: Statistical learning formulation of the DNA base-calling problem and its solution using a Bayesian EM framework. Discrete Appl. Math. 104, 229–258 (2000)

Semi-automated Mapping of Cell Nuclei in 3D-Stacks from Optical-Sectioning Microscopy

Martin Heß

Biozentrum der LMU München, Großhadernerstr. 2, 82152 Planegg, Germany
hess@zi.biologie.uni-muenchen.de

Abstract. 3D-stacks of optical sections through the vertebrate retina with fluorescent stained cell nuclei were measured with a laser scanning microscope. The evaluation of the data volumes with dedicated digital imaging algorithms gives access to complex morphometric tissue-characters that are discussed in terms of functional morphology. The thickness of nuclear layers and the 3D-coordinates of cell nuclei are detected automatically to measure cell densities, cell ratios and to create character-distribution-maps of the entire retina.

1 Introduction

Confocal laser scanning microscopy combined with any fluorescence staining technique is a powerful and elegant method to get three-dimensional structural data from biological tissues. Usually the result of a single xyz-scan-measurement is a stack of evenly spaced and perfectly aligned greyscale images ("optical sections") with a considerable data volume and information content. Frequently these stacks are used to generate attractive displays of the stained structures (e.g. brightest point projections, colour channel overlays, surface renderings), but rarely for thorough evaluation of the stack's information content by means of three-dimensional morphometric analysis. Many complex tissue characters are hardly revealed to an intuitive understanding by mere visual contemplation of 3D-data stacks or evade from a precise manual evaluation in an acceptable period of time. Digital imaging algorithms, however, allow the extraction of both simple and complex characters from huge data stacks in a time-saving way even on a standard PC. Still usually they have to be programmed and tailored to the specific object and scientific question by the scientist himself. Since biologists normally recoil from this challenge, most morphometric studies usually do not reach the third dimension to this day. In this study an example is presented for a computer-aided investigation and description of three-dimensional patterns of cell nuclei in the vertebrate retina. On the one hand retinal tissue is particularly suitable for optical-sectioning microscopy due to its transparency and low thickness, on the other hand the layered structure and high degree of geometrical order of this brain-derivative carries valuable information for the functional morphologist. For this study the retina of the European anchovy *Engraulis encrasicolus* (Teleostei, Engraulididae) was chosen to make a contribution to the morphometric description of the vertebrate retina in general and to approach to a more profound understanding of an uncommon retina in special, that is specialized for polarization contrast vision [1,2].

P. Perner and O. Salvetti (Eds.): MDA 2006/2007, LNAI 4826, pp. 156–164, 2007.

2 Material and Methods

2.1 Tissue Preparations

Adult European anchovies (*Engraulis encrasicolus*) where obtained from local fisherman just returning from their nocturnal catches in the Mediterranean (Adriatic sea, Rovinj). For the time of death dated back less than 1.5 hours, the retinal tissue of cooled animals could be regarded as *in-vivo*. Eyes where enucleated, the eyeballs perforated by razorblade-cuts through the cornea and fixed with 4% formaldehyde in 0.1M phosphate buffer at pH 7.4 plus 3% sucrose for several hours. The cornea, lens and vitreous body were removed in cold buffer, thereafter the entire retina of a right eye (diameter 8mm) was cut into 48 pieces whose original positions were documented. The retinal fragments were rinsed in buffer and embedded in 4% agarose at 45°C in separate dishes of two 24-well culture plates. From the centre of each fragment radial slices (thickness 50 µm) were made with a Leica VT1000S vibratome for subsequent radial optical sectioning. The slices were submersed in a 1µM-solution of TO-PRO-3 (Invitrogene, λ_{max}(Excitation) = 642nm, λ_{max}(Emission) = 660nm) in buffer for 10 to 60 minutes at ambient temperature for fluorescent staining of the cell nuclei. After the staining each slice was placed in a drop of anti-fading mounting medium (Vectashield®) between a glass slide and a cover glass and sealed with nail varnish. To avoid deformation of the slices by squeezing, the cover glass was braced by two pieces of cover glass as spacers (thickness 150µm) directly glued to the slide. A second preparation was accomplished to obtain 24 retinal fragments directly placed between 300µm spacers, pigment epithelium oriented downward for sub-sequent tangential optical sectioning.

2.2 Microscopy

The tissue preparations were imaged with a confocal laser scanning microscope (Leica TSC SP2 on an inverse Leica DM IRBE). For excitation of TO-PRO-3 the 633nm HeNe-line was used and attenuated to 10% to restrict bleaching. The beamsplitter was a triple dichroic (488, 568, 633nm) by default, the spectral detection window of the photomultiplier was set to 650-740nm. For the radial optical sections a Leica UV 25x PL Fluotar NA 0.75 oil objective was used (working distance 180µm, nominal resolution xy: 260nm, z: 1108nm), the voxel size was adjusted to 405nm in xy-plane by 810nm in z-direction (voxel-geometry: integral multiple of a cube). This allows both the display of radial slices through the thickest part of the retina in the "visual field" of the photomultiplier (207.4µm x 207.4µm allocated to 512 x 512 pixels) and a comfortable digital slice spacing without interpolation. Gain and offset of the photo-multiplier were optimized to exploit the 8bit-dynamic of the sensor with reference to the available signal. For the tangential optical sections a Leica UV 63x HCX PL Apo NA 1.32 oil objective was used, voxel size adjusted to 310nm in xy-Plane (158.7µm x 158.7µm allocated to 512 x 512 pixels) by 936nm in z-direction. The xyz-scans started near the cover glass towards the glass slide (against gravity), in each plane four optical sections were averaged to improve signal-to-noise ratio. Depending on the retina thickness and the tilt angle of nuclear layers in the tissue slices the number of optical slices was varied between 27 and 118. The resulting stacks of greyscale images had a data volume between 7.1 and 30.4 MB, altogether 0.9 TB of raw data were generated.

2.3 Digital Image Analysis

For further processing the image data stacks generated by the acquisition software of the confocal laser scanning microscope were imported in IDL (interactive data language, Research Systems Inc.) on a standard PC (2.7GHz, 1MB RAM) and subjected to several home-made IDL-algorithms. The line of actions - i.e. pre-processing, semi-automated detection of cell nuclei, mapping of measurements etc. - is subject of the results chapter.

3 Results

3.1 Data Import

3D-measurements at the CLSM usually deliver sequences of tiff-images as export-files. Every greyscale image can be regarded as a table of measurements sorted by columns and rows with entries between zero and 255 (8 bit). To get access to the entire data set of a 3D-measurement the respective image sequence was imported (via IDL software) into a single array-variable with three dimensions according to the x-, y- and z-axis of the measured tissue volume. The x- and y-index of the array mirrors the pixel-position in the original 2D-image, the z-index stands for the image number or its z-Position of the volume respectively. This allows to directly interrogate the measured value of the fluorescence signal of any point in the volume (voxel) specified by three index values. To get the correct proportions, every xy-plane was doubled (radial mechanical slices; z-spacing of optical sections 2x the pixel size) or trebled (tangentially oriented retina fragments; z-spacing 3x the pixel size). The last step can be omitted to save memory and to speed up calculations - for a correct display in perspectives (at any angle of view deviating from the z-axis) and for spatial measurements, however, the elongate voxel-shape has to be taken into consideration.

3.2 Display of Raw Data

In almost every case displays of the raw data show the nuclear layers of the retina oriented obliquely in the kartesian coordinate-system (Fig. 1). It is true that the vibratome sections were cut as close to the radial plane as possible under visual control, but a precisely radial orientation is not obtainable in practice, not least because of the hollow-sphere shape of the whole retina. Likewise in tangential view (whole mounts) the retina fragments always showed orientations tilted against the xy-plane. Nevertheless, as a simplification, a small retina fragment with 200 µm edges cut out of an eye with 8 mm diameter is regarded as not-curved in this study.

3.3 Cell Layer Alignement

To simplify the following measuring methods and to get the common depiction of the retina with horizontally aligned histological layers, the fluorescent stained nuclear horizons of the scanned retinal volumes are to be oriented as parallel to the xz-planes (radial sections) or the xy-planes (whole mounts) of the data volume as possible. This

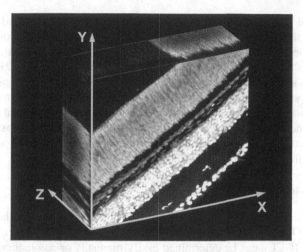

Fig. 1. Cubic display of a 3D-stack of optical slices through a small fragment of the anchovy-retina (radial vibratome slice) with coordinate system (X, Y, Z). The XY-view (207µm x 207µm) is a brightest-point-projection of the entire stack. The XZ- an YZ-views are single planes. *Note* nuclear layers lying obliquely in the data volume.

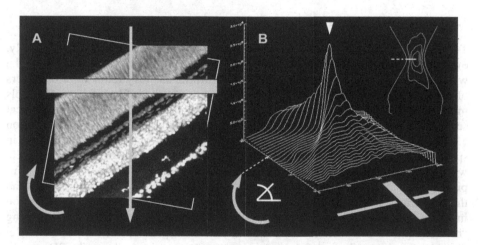

Fig. 2. Automated detection of rotation angle for the horizontal alignement of the nuclear layers. A) The XY-view of the data volume is convolved with a *horizontal bar* in vertical direction and then incrementally rotated. B) The resulting profiles build up a *2.5D-landscape* with a peak (*arrowhead*) that indicates the wanted rotation angle. *Inlay:* contour plot of the "mountain" with "summit"-position.

happened in two orthogonal directions either interactively (with auxiliary lines parallel to the x-, y- or z-axis) or automatically. An appropriate digital imaging algorithm is demonstrated on a radial 3D-scan exemplarily (Fig. 2, 3): a brightest point projection (BPP) of the data volume in the xy-plane (i.e. z-axis shortened to zero) serves to determine the inclination angle of the nuclear layers to the x-axis. To do this, the

BPP is rotated around the z-axis in 1° increments and convolved with a bright horizontal bar shifted vertically over the image in every angle-position. The result of this double-loop operation is a 2D-data set with a maximum indicating the rotation-angle that leads to horizontal alignement of the nuclear layers (Fig. 2). After rotation of the raw data stack around the z-axis by the determined angle (extension of the data volume on all sides helps to avoid clipping artefacts but increases memory demand and calculation time) the procedure is repeated with the yz-BPP and subsequent rotation around the x-axis. As a rule a second iteration of these two steps leads to a very good alignement of the nuclear layers for radial scans parallel to the xz-planes of the kartesian coordinate system (Fig. 3).

3.4   Simple Measurements

Based on BPPs of the aligned data volume the thickness of retinal layers, e.g. outer nuclear layer (ONL), inner nuclear layer (INL), inner plexiform layer (IPL) and ganglion cell layer (GCL), can be determined easily manually or automatically (Fig. 3). For the semi-automated morphometric analysis of the nuclear layers the definition of "regions of interest" (ROIs) containing unclipped fluorescence signals is required. The ROIs are defined on BPPs of the three orthogonal main-planes (XY, YZ, XZ), this way enclosing a "volume of interest" (VOI) completely filled with 3D-images of cell nuclei.

3.5   Detecting Nuclear Positions

To get the number and reliable centre-of-gravity positions of the cell nuclei quickly, every manually defined VOI was convolved with an idealized image (kernel) of the wanted structure (cell nucleus). The cell nuclei of the anchovy retina have diameters of 5-6μm depending on the cell type, resulting in circular profiles of 12 to 15 pixels maximally using the image acquisition settings indicated above for radial optical sections. Due to the almost spherical shape of cell nuclei in the retina the convolution can be executed with 2D-kernels plane by plane instead of 3D-kernels (spheres) in space. To do this, a kernel-array (2D-variable equivalent to the image of a white circle with the approx. nuclear diameter on a black background) is centred over each XY-pixel of the data volume subsequently. The overlapping pixel-values of the kernel and the image are multiplied and the result is stored at the corresponding centre-position in a new 3D-variable. The result of this convolution procedure is a data set containing "blurred light-clouds" with local maxima at the centre-positions of the wanted nuclei (Fig. 4). Starting with the brightest maximum of the entire volume, the 3D-coordinates of the corresponding nucleus was written into a table. Then the nucleus around the local maximum was deleted in the VOI by multiplication with a black sphere of the approx. nuclear diameter and the convolution was repeated with the modified VOI and so on. Stop-criterion for this procedure was an estimated and pre-defined number of iterations combined with a test for erroneous measurements. Starting from the first coordinate the distances to all other detected points in the VOI were calculated. If a value lower than twice the expected nuclear radius occurred, the relevant coordinate was deleted. Such "misdetections" accumulated at iteration numbers equal or larger than the actual number of cell nuclei in the VOI.

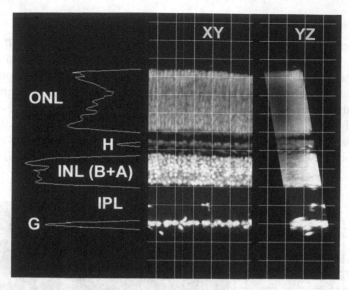

Fig. 3. Nuclear layers of the retina aligned parallel to the XZ-planes. Vertical intensity-profiles (*left*) help to measure the thickness of the outer nuclear layer (*ONL*), vitreal part of the inner nuclear layer (*INL* with bipolar (*B*) and amacrine (*A*) cells) and inner plexiform layer (*IPL*). Horizontal cells (*H*) and ganglion cells (*G*) form separate layers. *Note* restricted infiltration depth in the *ONL* (YZ-view).

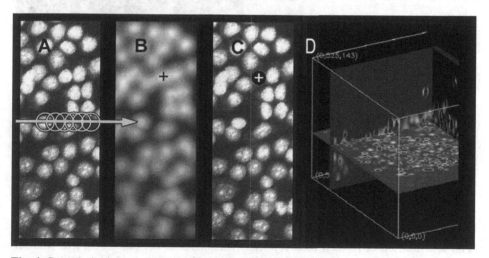

Fig. 4. Convolution of measured signals from the ganglion cell layer (*A*) with a discoidal kernel leads to a blurred picture with local maxima at the centre-positions of the cell nuclei (+ in *B*, 2D aspect of a 3D-operation). The detected nucleus is deleted (+ in *C*) prior to iteration of the convolution. *D*) Data stack with deleted nuclear centres.

3.6   Evaluation of Position Data

The corrected list of centre-coordinates allowed computation of the cell density of the VOI (converted to cells per $10^4 \mu m^2$ of retinal area), compilation of a neighbour-distances histogram for pattern-description (Fig. 5) and finally the correlation of measurements between the three neuron layers of a single retina fragment. After having analyzed several tissue fragments scattered over an entire retina 2D- or 3D-mapping of simple or complex measured characters can be demonstrated (e.g. density map of one cell type, ratio map of two parameters, Fig. 5). Every calculation step described above was performed on a standard PC between a few seconds and several minutes.

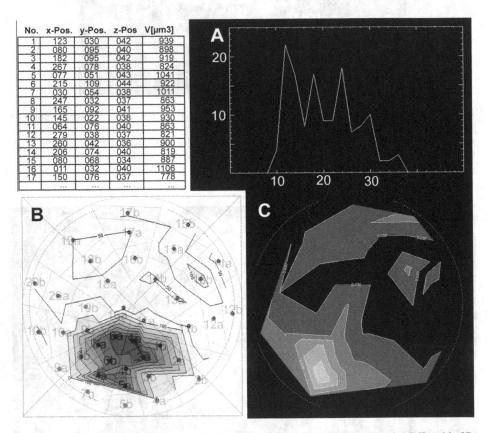

No.	x-Pos.	y-Pos.	z-Pos	V[µm3]
1	123	030	042	939
2	080	095	040	898
3	182	095	042	919
4	267	078	038	824
5	077	051	043	1041
6	215	109	044	922
7	030	054	038	1011
8	247	032	037	863
9	165	092	041	953
10	145	022	038	930
11	064	076	040	863
12	279	038	037	821
13	260	042	036	900
14	206	074	040	819
15	080	068	034	887
16	011	032	040	1106
17	150	076	037	778
...	...	...	...	...

Fig. 5. Evaluation of position data. *Top left*: Table of nuclei detected in the INL with 3D-coordinates and measured nuclear volume. *Top right*: Histogram of neighbour-distances in a small VOI of H-cells with peaks indicating a square pattern (x-axis in µm). *Bottom left*: density distribution of G-cells with a ventral maximum around 250 cells/$10^4 \mu m^2$. *Bottom right*: Ratio-map of INL-thickness / ONL-thickness indicating an area of high computing potential (light-grey) in the ventral retina.

4 Discussion

This study outlines a method that gives access to complex morphological tissue-characters arising from the spatial arrangement of cell nuclei (in the vertebrate retina as example) by the use of fluorescence staining, optical sectioning microscopy and digital image analysis algorithms tailored for special purposes. Additionally it is intended to demonstrate the usefulness of mass-data analysis in the field of histology and functional morphology and to encourage the ambitious life scientist to design his own application software. The study provides the following lessons and impulses:

4.1 Optical Sectioning Depth

Prior to any programming the image acquisition parameters have to be adapted or rather optimized to get data sets that are suitable for evaluation. To be able to excite and collect fluorescence light from a sufficient tissue volume a penetration depth of at least 50 µm is desirable. Despite a relative high optical transparency, formalin-fixed retina tissue considerably scatters the visible light inversely proportional to its wave-length. To get deep optical sections with a satisfactory signal-to-noise-ratio a fluores-cent stain with excitation- and emission maxima in the "red part" of the electro-magnetic spectrum should be favoured (e.g. TO-PRO-3). The limiting factor in terms of penetration depth turned out to be the restricted infiltration of the ONL by different dyes even with infiltration times of more than 1h at 30° (see YZ-view in Fig. 3). As not much more than 25 µm of the tissue can be stained in z-direction the thickness of mechanical radial sections should not exceed 50 µm in this case.

4.2 Field-of-View and Resolution

The use of a red fluorescent dye happens somewhat at the expense of spatial resolution but doesn't influence the conspicuousness of cell nuclei. In fact the field-of-view of the microscope's sensory device has to be adjusted to gather fluorescent light from all three nuclear layers of the retina at the same time (radial slices) to be able to correlate cell counts of radial staggered VOIs and to minimize the total data volume. In the examined material the distance between the vitreal border of the GCL and the scleral border of the ONL peaked at about 230 µm, fitting diagonally in the chosen field-of-view. The resulting nuclear diameters of around 15 pixels turn out to be an acceptable trade-off. Tangential optical sections of mechanically not-sectioned retina fragments were made to record VOIs of the GCL (radial scans with a z-size of ≤ 50 µm were less suitable to image this nuclear monolayer). In this orientation the resolution was increased by the factor of 1.3, the field-of-view restricted respectively.

4.3 3D-Arrays

The import of image stacks into a single 3D-array in general and the use of IDL in particular allows comfortable access to every single voxel-value and to easily apply a series of powerful imaging-routines and other logical operations. Programming in a compiler language like IDL opens up the possibility to compute large data sets and frequent iterations even on standard PCs and notebooks relatively fast – ImageJ for example does the same job in a comparatively unacceptably long period of time. Of course similar approaches were developed and conducted also by other investigators for their special research problems independently [3].

4.4 Functional Morphology

From the zoological perspective the semi-automated analysis of nuclear patterns opens the door to the description and interpretation of tissue-characters (not only in the retina) that hitherto has been too time-consuming or even impossible with paper and pencil. This is especially true for an accurate counting of objects (e.g. cell nuclei) not only on single microscope-slides [4], but also in high-content data stacks, also for generating neighbour-distance histograms in 3D (to describe cellular sphere-packing patterns or developmental processes at the retinal margin) and for the display of standardised density- and ratio-maps. Some examples are given to illustrate the scope of the functional morphological discussion of complex retinal characters: The density distribution of photoreceptors in the retina gives an indication of the visual acuity in different sectors of the visual field. This is, however, only reliable if the density distribution of ganglion cells mirrors the first mentioned pattern. The ratio of photoreceptors to ganglion cells in a small area reveals the degree of radial signal convergence and thus an relative indicator of light sensitivity (cone- and rod-pathways have to be analysed separately), the ratio of photoreceptors to secondary neurons, on the other hand, gives indications about the potential computing power or computing complexity of the examined retina fragment etc [5].

4.5 Outlook

To continue with this subject it is planned to expand the image analysis applications to the automated recording of shape-parameters and the high-resolution distribution of the fluorescence signal within single nuclei for cell-classification (in combination with neuroanatomical techniques), to the mapping of nuclei in hemispheric coordinate systems (e.g. small eyes) and ultimately to the full-automated adaptive acquisition and analysis of fluorescence signals from entire retinae with motorized microscopes. High-content applications would be also the comparison of developmental stages or related species.

References

1. Fineran, B.A., Nicol, J.A.C.: Studies on the photoreceptors of *Anchoa mitchilli* and *A. hepsetus* (Engraulidae). Phil. Trans. R. Soc. Lond. B 283, 25–60 (1978)
2. Heß, M., Melzer, R.R., Smola, U.: The Pattern of cone pedicles and horizontal cells in the retina of the European anchovy *Engraulis encrasicolus* L (Engraulididae, Clupeiformes). J. Submicroscopic Cytol. Pathol. 34(4), 355–365 (2002)
3. Bredno, J., Metzler, V., Nacimiento, W., Lehmann, T.M., Spiter, K.: Detektion und Quatifizierung der Membranstrukturen von Nervenzellen. In: Lehmann, T.M., Metzler, V., Spitzer, K., Tolxdorff, T. (eds.) Bildverarbeitung für die Medizin, pp. 407–411. Springer, Berlin (1998)
4. O'Connell, C.P.: The structure of the eye of *Sardinops caerulea*, *Engraulis mordax* and four other pelagic marine teleosts. J. Morph. 113, 287–329 (1963)
5. Archer, S.N., Djamgoz, M.B.A., Loew, E.R., Partridge, J.C., Vallerga, S.: Adaptive mechanisms in the ecology of vision, p. 668. Kluwer Academic Publishers, Dordrecht, Boston, London (1999)

Statistical Analysis of Myocyte Orientations of the Left Ventricular Myocardium

Kai Rothaus and Xiaoyi Jiang

Department of Computer Science, University of Münster
Einsteinstrasse 62, D-48149 Münster, Germany
{rothaus,xjiang}@math.uni-muenster.de

Abstract. The commonly used model of the heart for medical applications suffers from some incompleteness when explaining different kinds of measured forces in vivo studies by medical experts. In this paper, we make a statistical analysis of the so-called angle of intrusion automatically. The basis of the proposed method is a set of histological preparations showing heart fibre tissue. We adapt a multi-scale midline extraction process to extract the myocyte strings out of these images and measure the angles of intrusion. Furthermore, a statistical model is derived and validated by the result of a novel parameter estimation technique.

1 Introduction

In this work we present an approach to analysing the orientation of myocyte strings automatically. For this, digitised images showing heart tissue (see Figure 1 for an example) are processed. The dark, elongated structures represent the strings of myocyte cells, which cause the contraction of the heart muscle. The top border of the image is oriented parallel to the epicard (the outer border of the heart). Substantially, the myocyte cells form strings, which are situated parallel to the epicard with slight variations. For medical purpose, the distribution of the myocyte orientations, also denoted as angle of intrusion, has a high impact. Using former models of the heart, where the myocyte structures are essentially ignored, the forces observed by the physicians in vivo studies [5] cannot be simulated or at least explained suitably. Lunkenheimer et al. [6] try to enhance the existing model of the heart by investigating two different kinds of observed forces. Their assumption is that there must be not only tangential directed myocyte strings, as assumed so far. Furthermore, they expect a larger portion of transversal myocyte strings. The first step to document this assumption is an appropriate analysis of the angles of intrusion, which we present in this paper.

Some former work has been done to perform quantitative assessments of myocytes by Karlon et al. [2]. They compare manual measurements with two automatic approaches. The first method is based on a Hough transform technique. Firstly, edges are computed using four different gradient masks. The responses of these filters are thresholded and a connected component analysis is performed. Afterwards, the image is divided into smaller regions, and some constraints are checked to filter out false regions. On all remaining regions a Hough transform is performed to compute one mean

P. Perner and O. Salvetti (Eds.): MDA 2006/2007, LNAI 4826, pp. 165–175, 2007.

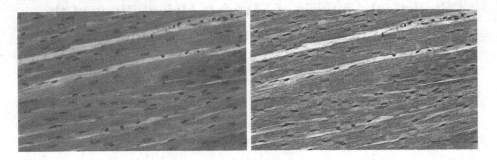

Fig. 1. Sample image of heart tissue: Myocyte **Fig. 2.** Grey-level transform and adaptive con-
strings are visible as dark elongated structures   trast enhancement of the slice in Figure 2

orientation of the structures in each. The mean orientations of all regions are collected
and constitute the observation set.

The second method described by Karlon et al. [2] is based on the intensity image
gradient directly. Once again, the image is divided into regions and a statistical analysis
is performed in each. As statistical model, the class of von Mises distributions with
parameter kappa (for statistical background see [1]) is used. Since the result of the
two methods are justified by a manual analysis, an automatic analysis of the angle of
intrusion is well founded [2].

In contrast to these two methods, our approach does not divide the images into regions. Since we are not only interested in the mean angle of intrusion, but also want to
analyse the underlying distribution, we try to take as much information as possible into
consideration. For this, we locate the midlines of the myocyte strings in the images and
take the tangential vectors at sample points of the extracted midlines as observations.

The remainder of this paper is organised as follows. At first, we describe the image
analysis part (Section 2). This process results in the observation set consisting of local
measured angles of intrusion. In Section 3 we give the statistical analysis of the observation set. We will present two different distribution models, based on classes of von
Mises and Gaussian distributions, respectively. Afterwards, the results of our approach
are shown (Section 4). Finally, we end up by drawing some conclusions (Section 5).

2   Analysis of Heart Fibre Images

The histological preparations are cuts of pig hearts, which are dissected using pairs of
cylindrical knives with different diameters (see [6]). After pinning the slices flat, they
are fixed in formaldehyde, embedded in paraffin and sectioned. A treatment with several
substances is done to achieve a swelling of the preparations, so that the myocyte strings
are clearly visible. The colour images show these preparations 100 times magnified. All
slices are adjusted with the upper border parallel to the epicard and recorded on digital
camara.

The image analysis part of our method consists of three steps: enhancement of the myocyte strings (Section 2.1), midline extraction (Section 2.2) and measurements of the tangential orientations of the midlines at equidistant sample points (Section 2.3).

2.1 Image Acquisition and Enhancement of the Myocyte Strings

At first, we transform the colour images to intensity images $I(x, y)$ by a linear combination of the three RGB colour channels.

$$I(x, y) = 0.2626 \cdot R(x, y) + 0.4116 \cdot G(x, y) + 0.3258 \cdot B(x, y) \qquad (1)$$

This linear combination scheme is computed once by analysing a subset of images by a principal component analysis, to keep as much contrast in the images as possible.

Afterwards, we enhance the contrast of the image $I(x, y)$ by an adaptation of the method proposed by Yu et al. [9]. This method works by propagating the minimum ($lmin$), average ($lavg$) and maximum ($lmax$) value towards different scan directions by a conditional propagation scheme. The initialisation of the three arrays $lmin$, $lavg$ and $lmax$ is the image, which should be processed. In contrast to the original approach of Yu et al. [9] we use different conductivity factors for the minimum ($C_{min} = 0.95$), average ($C_{avg} = 0.75$) and maximum ($C_{max} = 0.55$) values, to steer the propagation behaviour of the enhancement algorithm. Thus, we use the conditional following propagation schemes, to update the local image features at the actual scanned position:

$$lavg \leftarrow (1 - C_{avg}) \cdot lavg + C_{avg} \cdot \overline{lavg} \qquad (2)$$

$$lmin \leftarrow (1 - C_{min}) \cdot lmin + C_{min} \cdot \overline{lmin} \qquad \text{iff } lmin > \overline{lmin} \qquad (3)$$

$$lmax \leftarrow (1 - C_{max}) \cdot lmax + C_{max} \cdot \overline{lmax} \qquad \text{iff } lmax < \overline{lmax} \qquad (4)$$

where the bar denotes the value at the previous scanned image position. Furthermore, we choose two scan directions, namely from top to bottom and vice versa. This adaptation of the approach is motivated by our goal to keep the dark structures, but lighten the bright structures to enhance the contrast. For each pixel p the three resulting values $lmin, lavg, lmax \in [0, 1]$ reflect the minimum, average and the maximum intensity in a neighbourhood of p. The original intensity old at p can now be emphasised against its neighbourhood. Therefore, we define the local intensity range as $\delta = lmax - lmin$ and the local enhancement factor as $\omega = \sqrt{\delta \cdot (2 - \delta)}$. The new intensity value at p is then computed as (adaption of the transformation proposed by Yu et al. [9]).

$$\frac{lmax + lmin - \omega}{2} + \frac{2\,\omega\,(old - lmin)\,(\delta + \omega\,(lavg - old)\,(old - lmax))}{\delta^3}. \qquad (5)$$

The result of this grey level transform and enhancement step applied on the preparation of Figure 1 is shown in Figure 2.

2.2 Midline Extraction

After the pre-processing procedure (Section 2.1), now the extraction of the midlines will be explained. For this task we have developed a multi-scale extension [7] of López's

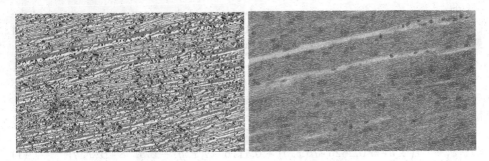

Fig. 3. Gradient image after multi-scale smooth- **Fig. 4.** Extracted midlines are laid over the in-
ing of the gradient vector field   tensity image

Level-Set-Extrinsic-Curvature approach (LSEC) [3,4]. In the following, we give a brief summary of this extension. The pixel array I of the enhanced intensity image (Figure 2) is taken as input image, in which we want to localise the midlines of dark elongated structures (myocyte strings).

(1) Computation of the intensity gradient vector field:
After a slight smoothing of I with a Gaussian kernel ($\sigma = 0.75$) we apply the Sobel operator, which results in the partial deviation I_x and I_y. These deviations are used to compute the edge magnitude array S and a local edge orientation array Θ. Since at each pixel, the corresponding deviations I_x and I_y could be recomputed on S and Θ, in the following we use the notation (I_x, I_y) or (S, Θ) for the gradient vector field equivalently.

(2) Enhancement of edge magnitude:
Since the LSEC approach of López et al. [3,4] is valid on normalised gradient images only, we boost the edge magnitudes S pixel-wise using the function

$$b_t(S) = 1 - \exp\left(-\frac{S^2}{2 \cdot t^2}\right) \tag{6}$$

with threshold $t = 0.075$. This optimistically choice is made to preserve even weak edges, since they would be filtered out in the further process if there are no equally orientated edges in the neighbourhood.

(3) Iterative smoothing process:
The goal of this step is to smooth the gradient vector field in the sense that the gradient vectors are propagated towards the interior of dark elongate structures. Firstly, the structured tensor is computed for each pixel

$$ST(x, y) = \begin{pmatrix} I_x(x,y)^2 & I_x(x,y) \cdot I_y(x,y) \\ I_x(x,y) \cdot I_y(x,y) & I_y(x,y)^2 \end{pmatrix}, \tag{7}$$

where (x, y) are the image coordinates. This tensor field is smoothed element-wise with Gaussian kernels G_k of different scales $\sigma_k = k \cdot \sigma_0$ ($\sigma_0 = 0.75$). For each pixel (x, y) and each scale we compute the edge magnitude $S_k(x, y)$ and the new edge orientation $\Theta_k(x, y)$ based on the largest eigenvalue and corresponding eigenvector of the smoothed structured tensor (for details see [7]).

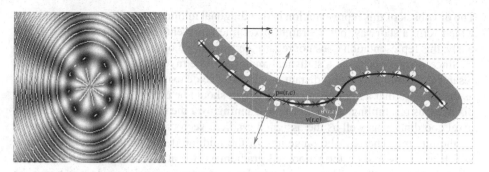

Fig. 5. Example image of López   **Fig. 6.** Equidistant resampling of the midlines and computing of
et al.[3]   the angle of intrusion

Finally, for the pixel (x, y) the vector with the highest edge magnitude value $S_k(x, y)$ of all considered scales k is chosen as the result vector of the iterative smoothing process. In Figure 3 the result of the sample image (Figure 1) is shown. The magnitude S of the vectors are visualised by the intensity (V) and the direction Θ by the colour (H) in the HSV colour space.

(4) Computation of local creaseness in the intensity image: After a single additional smoothing, we apply the divergence operator on the smoothed gradient vector field (I_x, I_y). The level set extrinsic curvature is a creaseness measure of an image function I. The midlines we want to detect consist of pixels with maximum creaseness. Therefore, the level set extrinsic curvature is computed as the negative divergence κ of the smoothed gradient at each pixel. López et al. [3,4] have proved, that this is equivalent to the direct computation

$$\kappa(x,y) = \frac{2I_x(x,y)I_y(x,y)I_{xy}(x,y) - I_y^2(x,y)I_{xx}(x,y) - I_x^2(x,y)I_{yy}(x,y)}{\left(I_x^2(x,y) + I_y^2(x,y)\right)^{3/2}}s \quad (8)$$

of the LSEC in continuous domains under certain preconditions. In fact, the divergence of gradients gives even better results in discrete domains than the direct computation. This advantage is depicted in the example image of Figure 5. The divergence approach leads to continuous segments, where as the direct computation leads to gaps [4].

(5) Grouping points of maximal creaseness to line segments: Pixels, which hold a local maximum creaseness, value in direction of the local smoothed gradient vector are taken as candidates for midline pixels. We link two neighboured candidates together if the gradients and the creaseness at the corresponding pixel are similar. This can be done by simple threshold rules. After this grouping process and an additional filtering step (discard segments of less than three pixels), we get the midline segments, which represent the myocyte strings. Result of this step are presented in the Figures 4 and 5, where the extracted midline segments are overlaid on the grey-scale images.

2.3 Measurement of the Tangential Orientations

The midlines are represented as strings of neighboured midline pixels (see Figure 6). With each midline pixel (dots) the smoothed gradient vector is stored (connected arrows).

The drawback of this representation is that diagonal midlines are represented by pixels, which form a stairway, but a straight pixel line represents horizontal midlines. Obviously, this representation does not regard the true length of the midline.

Since we want to make a statistical analysis of the myocyte orientation, we have to resample the midlines at equidistant points. We decide to choose the width of one pixel as distance, so that a midline of n pixel length should be represented by $n + 1$ midline pixels. This resampling can be done by scanning over the midline and interpolating the pixel coordinates as well as the assigned gradient vectors using the nearest two pixels on the midline. Thereby, the tangential of the midline defines the scanning direction, which is orthogonal on the smoothed gradient vector at the last considered midline point. In Figure 6 the resulting pixel coordinates (crosses) are shown. In this figure the computation of the tangential vector v at pixel p with coordinates (r, c) is drafted.

Naturally, the situation in Figure 6 is idealised, but in fact we are mainly interested in the analysis of the tangential vectors. For the purpose of midline visualisation, we keep the representation at pixel grid points, whereas for the statistical analysis we choose the equidistant representation with sub-pixel accuracy.

3   Statistical Data Analysis

The tangential vectors at the equidistantly distributed sample points of the myocyte string midline are taken as observation set. Obviously, these vectors are represented in a cyclic domain with period of $180°$ or π, respectively. For this reason, we have to derive a model for cyclic data spaces. Fisher [1] gives a general introduction in the statistical analysis of such data spaces. In the following, we treat each observed angle as a point on a circle. Since the period of the domain is $180°$, an angle does not correspond to the normal angle in the Euclidian manner. For the purpose of geometrical interpretations of the observation set in the cyclic domain, all angles have to be multiplied by 2.0.

One can observe three common characteristics of the observed angles of intrusions:

1. Presence of equally distributed noise, introduced by falsely detected structures.
2. The angles are unimodal distributed (see Figure 7) with only a slight variation.
3. The distribution seems to be symmetrical.

Based on these observations, we have derived a suitable model for the underlying distributions. Due to the first characteristic, the noise is regarded as an additive constant term α/π, where α is the portion of noise. To model the signal (i.e. the angle of intrusion) a symmetrical density function with one local maximum should be used. In Section 3.1 we present two models (Gaussian and von Mises, respectively). The parameter estimation procedure is the same for both models. Since the equally distributed noise has no impact to the calculation of the mean orientation, at first the mean orientation $\hat{\mu}$ is estimated. We have tested several estimators [8] and figured out that a least median error approach works best (Section 3.2). Subsequently, the portion of noise is estimated by inspecting the neighbourhood of the antipole of $\hat{\mu}$ (Section 3.3). At last we compute the estimation of the shape parameter (Section 3.4), which are $\hat{\kappa}$ (von Mises) and $\hat{\sigma}$ (Gaussian), respectively.

3.1 Distribution Models

Both distribution models, which we have taken into considerations, are explained by three parameters: the portion of noise $\alpha \in [0, 1]$, the mean angle $\mu \in [-\pi/2, \pi/2)$, and one shape parameter. In detail, the PDFs are given as:

$$p_{\alpha,\mu,\sigma}^{\text{Gauss}}(\phi) = \frac{\alpha}{\pi} + \frac{(1-\alpha) \cdot \exp(-\frac{(2\phi-2\mu)^2}{2\sigma^2})}{\pi\,\sigma} \tag{9}$$

$$p_{\alpha,\mu,\kappa}^{\text{von Mises}}(\phi) = \frac{\alpha}{\pi} + \frac{(1-\alpha) \cdot \exp(\kappa\,\cos(2\phi - 2\mu))}{\pi\,I_0(\kappa)} \tag{10}$$

where I_0 is the modified Bessel function of order zero. The first summand of these density functions represents the equally distributed portion of noise with fraction α and the second summand (with fraction $1 - \alpha$) is an adaptation of the signal distribution to the cyclic domain of $\phi \in [-\pi/2, \pi/2)$.

The Gaussian model cannot be applied directly to the observation set, due to the cyclicity of the data space. For this, we have to cut the data space at the antipole of the mean angle. Afterwards, the whole observation is mapped to $\mathbb{R}$ linearly, such that the mean value is mapped to 0 and finally, a standard analysis on the line is performed to estimate the standard deviation σ of the underlying Gaussian.

The use of a von Mises model offers the evidence that the shape parameter κ is directly deducible on the cyclic domain. Note that a reduction of κ leads to an enlargement of the variance. The von Mises distribution is a kind of standard distribution on cyclic domains (see [1]). Its importance is comparable with the role of the normal distribution on a line. Unfortunately, things becomes more difficult on cyclic domains, so that not all properties of the normal distribution are adaptable to von Mises distributions. The close relationship becomes clear, by inspecting the density function of the von Mises model. If the cosine in Equation (10) is approximated by its first ordered Taylor polynomial the model is (beside normation and $\kappa := \sigma^{-2}$) the same as the Gaussian (see Eq. (9)). However, in all conducted experiments the von Mises distribution seems to model the observation more precisely (Table 1).

For validating the goodness of our models and to compare different estimators, we have used the following quality measure, denoted as match score (MS). The data space is divided into finite intervals of equal length (we use 180 bins $b_1, \ldots b_{180}$ of $1°$ width as default). For this discretisation we construct the relative histogram h of the observed angles of intrusion and compare for each bin b_i the measured frequency $h(i)$ with the expected frequency $f(i)$. Assuming a distribution with parameter vector p the match score is computed as:

$$\text{MS}(p) = 1 - 0.5 \cdot \sum_{i=1}^{180} |f(i) - h(i)|. \tag{11}$$

3.2 LME-Estimation of the Mean Orientation

For a fixed μ the median error to the observation set $S = \{\phi_1, \ldots, \phi_n\}$ is computed as

$$\text{mederr}(\mu, S) = \text{median}_{i \in \{1,\ldots,n\}} \{ d_{\text{arc}}(\mu, \phi_i) \} \tag{12}$$

where d_{arc} is the arc-length distance in our cyclic domain. Now, we minimise this error function err(μ) to estimate μ

$$\text{err}(\mu) = \min_{\mu \in [-\pi/2, +\pi/2]} \{ \text{mederr}(\mu, S) \}, \tag{13}$$

$$\hat{\mu} = \arg \min_{\mu \in [-\pi/2, +\pi/2]} \{ \text{mederr}(\mu, S) \}. \tag{14}$$

The statistical interpretation of this method is minimising the width of the 50%-quantile interval centred at μ.

The computation of $\hat{\mu}$ can be done by a single scan through the sorted observation set $\phi_{(1)}, \ldots, \phi_{(n)}$. We use two cyclic indices i_A and $i_\Omega = i_A + n/2$, which represent the beginning and the ending of a candidate error interval. For each error interval the corresponding estimation for μ is given by $(\phi_{(i_A)} + \phi_{(i_\Omega)})/2$. The estimation $\hat{\mu}$ is computed as the centre of the shortest candidate error interval and can be found by one sweep.

3.3 Estimating the Portion of Noise

After estimating the mean direction $\hat{\mu}$, we utilise the characteristic that the variation of the data is less in comparison to the domain. Therefore, we assume that all observed angles in a small interval centred at the antipole $\tilde{\mu} := \hat{\mu} + \pi/2$ of $\hat{\mu}$ are introduced by noise. In our implementation, we choose an interval length of $0.2 \cdot \pi$, which gives appropriate results. Since we assume equally distributed noise on the whole domain, we can easily estimate the overall portion of noise: Let n be the size of the observation set and c be the counted angles in the interval $]\tilde{\mu} - 0.1\pi, \tilde{\mu} + 0.1\pi]$ the portion of noise could be estimated as $\hat{\alpha} = 5c/n$.

3.4 Estimating the Shape Parameter

Finally, we estimate the shape parameter, which is κ in the von Mises case. This can been done by inspecting the minimised error $err := \text{err}(\hat{\mu})$ (see Eq. (13)). We compute $\hat{\kappa}$ as the unique solution of the equation

$$0.5 = \int_{-err}^{err} p_{\hat{\mu}, \hat{\kappa}, \hat{\alpha}}(\phi) \, d\phi. \tag{15}$$

Unfortunately, this equation has no analytical solution. We found κ by using the fact that the right side of the objective functional is increasing in $\hat{\kappa}$, so that we can guarantee to find κ by a binary search algorithm. This estimation technique can be adapted to other distributions, which have one shape parameter (e.g. the Gaussian model).

4   Results

Our image database consists of 45 images showing heart tissue. All slices are arranged in the common coordinate system, such that the epicard is parallel to the upper border of the image. We have estimated the parameter of the proposed model for all images and compute the match score (see Table 1). Furthermore, the average of the single results

Table 1. Results of the statistical analysis on the whole image data base

Preparation	Size	$\hat{\mu}$	$\hat{\kappa}$	$\hat{\alpha}$	MS von Mises	MS Gaussian
Slice 01	57728	-3.08	10.80	2.97%	91.24%	91.09%
Slice 02	54608	10.07	7.61	4.07%	91.05%	90.58%
Slice 03	61283	7.42	9.08	3.26%	92.40%	92.00%
Slice 04	69057	16.72	12.97	1.77%	91.62%	91.36%
Slice 05	60890	17.83	7.35	2.94%	93.29%	92.83%
Slice 06	58330	-3.44	5.86	7.02%	92.54%	91.88%
Slice 07	66060	3.69	8.66	3.84%	92.40%	91.96%
Slice 08	61291	-1.79	7.30	5.42%	92.33%	91.85%
Slice 09	55559	-11.93	4.98	7.50%	93.25%	92.56%
Slice 10	60250	0.15	7.53	6.43%	91.85%	91.39%
Slice 11	56364	-5.46	4.82	11.34%	92.09%	91.31%
Slice 12	60266	-1.22	4.70	12.05%	92.00%	91.20%
Slice 13	56292	4.78	4.74	6.47%	93.31%	92.45%
Slice 14	57900	-0.01	5.41	7.38%	90.84%	90.16%
Slice 15	54972	6.40	4.14	8.53%	92.58%	91.61%
Slice 16	64023	-3.82	7.45	5.40%	90.29%	89.80%
Slice 17	58538	-7.73	4.15	5.55%	93.72%	92.91%
Slice 18	60359	-0.43	5.79	6.20%	93.01%	92.39%
Slice 19	62503	2.18	8.38	3.18%	93.61%	93.22%
Slice 20	59616	0.09	5.21	4.34%	92.45%	91.76%
Slice 21	60453	-1.60	5.47	8.92%	91.74%	91.08%
Slice 22	60085	3.74	6.77	6.92%	91.59%	91.03%
Slice 23	61623	-5.41	3.76	10.73%	93.11%	92.10%
Slice 24	60643	-1.25	4.03	4.37%	93.34%	92.36%
Slice 25	61079	2.63	5.97	6.20%	92.62%	91.96%
Slice 26	63194	17.74	5.15	8.37%	92.85%	92.13%
Slice 27	58143	1.53	4.29	10.27%	92.62%	91.75%
Slice 28	59888	2.48	9.66	6.13%	91.29%	90.93%
Slice 29	60537	1.30	4.62	12.31%	90.92%	90.25%
Slice 30	57753	5.82	5.39	9.51%	89.81%	89.10%
Slice 31	56389	0.76	6.30	12.16%	91.54%	91.00%
Slice 32	54771	0.55	6.28	8.90%	90.59%	89.98%
Slice 33	64169	-2.17	6.50	5.71%	91.05%	90.51%
Slice 34	66315	-0.72	4.47	4.62%	92.04%	91.20%
Slice 35	54963	-11.74	5.44	8.15%	93.65%	93.00%
Slice 36	58668	-0.10	4.00	7.74%	92.70%	91.74%
Slice 37	63878	-8.31	6.11	6.46%	90.94%	90.39%
Slice 38	58285	-1.47	7.20	4.90%	92.07%	91.63%
Slice 39	56380	-0.88	4.53	7.05%	93.42%	92.63%
Slice 40	54885	-0.35	6.16	10.06%	92.53%	91.99%
Slice 41	60458	-12.58	5.44	7.05%	93.86%	93.20%
Slice 42	53136	1.31	3.29	12.62%	92.33%	91.17%
Slice 43	47940	4.96	8.58	6.39%	93.73%	93.34%
Slice 44	62925	17.67	10.12	2.33%	91.56%	91.21%
Slice 45	66382	8.60	13.71	2.33%	90.78%	90.51%
Average	59530	1.18	6.45	6.80%	92.19%	91.57%
Complete Set	2678831	1.17	4.27	6.75%	94.87%	93.98%

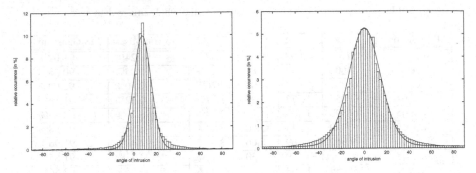

Fig. 7. Angle of intrusion histogram for the example in Figure 1

Fig. 8. Angle of intrusion histogram collected in 45 slices

are plotted in the row 'Average' and the analysis of the whole observation set, which consists of the union of all observation sets are presented in the row 'Complete Set'. This combination is valid because all slices are referenced to the same context.

The first observation is that in every case the von Mises model outperforms the Gaussian model by mostly over 0.5 match score points. The average match score of the von Mises model is with 92.19% reasonable high (98.24% in the case of simulated data) and validates our model. In Figure 7 we have plotted the observation histogram against the derived PDF (normalised to the same integration area) of the sample slice. The approximation of the model with the observation is good, except near the mean orientation. In some other slices, we obtain local marginal inaccuracies other where, but not at a fixed position. This phenomenon could be explained with local differences in the heart tissue, which exhibit different structures at different locations of the heart muscle. Since the preparations in our test study are located all over the heart, combining all results eliminate local phenomena. In Figure 8 the observation histogram of the aggregated observations and the derived PDF are overlaid. The visible model-observation coherency in this plot demonstrates the overall goodness of our model. This is also confirmed by the match score indicator of the complete evaluation of 94.87%. Here, the match score 93.98% of the Gaussian estimation is clearly lower.

The estimations of μ and κ correspond to the expectancy of the medical experts and in their opinion confirm the need of a more accurate model of the heart [6].

5   Conclusions and Further Work

In this paper, we have presented a completely automatic method for analysing the angle of intrusion of myocyte strings in heart tissue slices. Firstly, we have described an algorithm to extract the myocyte string and then given a method to measure the angle of intrusion at a multitude of sample points. Furthermore, we have developed a statistical model for the angle of intrusion distribution and validated this model experimentally.

Motivated by the results we want to advance the improvement of the heart model, by extracting the structure of the myocyte strings, which are connected to each other.

Moreover, we are discussing with other research groups if a simulation of a heartbeat based on such image material could be possible.

References

1. Fisher, N.I.: Statistical analysis of circular data. Cambridge University Press, Cambridge (1993)
2. Karlon, W.J., Covell, J.W., McCulloch, A.D., Hunter, J.J., Omens, J.H.: Automated measurement of myofiber disarray in transgenic mice with ventricular erxpression of ras. The Anatomical Record 252(4), 612–625 (1998)
3. López, A.M., Lumbreras, F., Serrat, J.: Creaseness from level set extrinsic curvature. In: Burkhardt, H., Neumann, B. (eds.) ECCV 1998. LNCS, vol. 1407, pp. 156–169. Springer, Heidelberg (1998)
4. López, A.M., Lumbreras, F., Serrat, J., Villanueva, J.: Evaluation of methods for ridge and valley detection. IEEE Trans. on PAMI 21(4), 327–335 (1999)
5. Lunkenheimer, P.P., Redmann, K., Florek, J., Fassnacht, U., Cryer, C.W., Wübbeling, F., Niederer, P., Anderson, R.H.: The forces generated within the musculature of the left ventricular wall. Heart 90, 200–207 (2004)
6. Lunkenheimer, P.P., Redmann, K., Kling, N., Jiang, X., Rothaus, K., Cryer, C.W., Wübbeling, F., Niederer, P., Heitz, P.U., Ho, S.Y., Anderson, R.H.: Three-dimensional architecture of the left ventricular myocardium. Anat. Rec. 288A(6), 565–578 (2006)
7. Rothaus, K., Jiang, X.: Multi-scale midline extraction using creaseness. In: Singh, S., Singh, M., Apte, C., Perner, P. (eds.) ICAPR 2005. LNCS, vol. 3687, pp. 502–511. Springer, Heidelberg (2005)
8. Rothaus, K., Jiang, X.: Comparison of methods for hyperspherical data averaging and parameter estimation. In: Proc. of the ICPR, vol. 3, pp. 395–399 (August 2006)
9. Yu, Z., Bajaj, C.L.: A fast and adaptive method for image contrast enhancement. In: Proc. of the ICIP, vol. 2, pp. 1001–1004 (October 2004)

Automatic Quality Control of Cereals in Particular Wheat-Subtask Detection of Hygiene-Relevant Parameters

Petra Perner and Thomas Günther*

Institute of Computer Vision and Applied Computer Sciences, IBaI
04107 Leipzig, Körnerstr. 10
* JenaBios GmbH, D-07749 Jena, Loebstedter Str. 78
pperner@ibai-institut.de
www.ibai-institut.de

Abstract. We are going on to develop a novel method for the detection of hygiene-relevant parameters from grains of cereal crops based on intelligent image acquisition and interpretation methods as well as data mining methods. The work presented here is part of a larger project aiming to develop an automatic system to the determination of the quality of cereals in particular wheat. We present our first results that describe the data acquisition, the planned image analysis and interpretation method as well as the reasoning methods that can map the automatic acquired parameters of grain to the relevant hygiene parameters. The preliminary results show that with the new computer science methods it is possible to come up with new insights into the quality control of food stuff.

1 Introduction

Fungal contamination of cereals is a serious economic problem throughout the world. Several fungi cause a reduction of grain quality, especially changes in color and taste [Müller et al., 1997], [Herrman et al, 1998], and [Rodeman, 2003]. However the main riks of fungal damage arise from the production of toxic compounds, known as mycotoxins. Mycotoxins can cause serious adverse health effects. Toxigenic fungi that produce mycotoxins in grains of cereals or oil seeds belong to the genera Aspergillus, Alternaria, Fusarium and Penicillium. The control of this problem is therefore of particularly interest in food safety and quality control programs.

The work presented here is part of a larger project aiming to develop an automatic system to the determination of the quality of cereals in particular wheat. The aim of the research is the development of an automatic system to the determination of the quality of cereals in particular wheat. A subtask of the system is the fast recognition of cereal grains damaged by fungi. Thereby should be developed a data acquisition unit that allows taking the coverage from the grain and allows placing it under a microscope for the acquisition of a digital image. This image should be used in order to automatically determine the number and the kind of fungi spores contained on the grain. For that we have to develop suitable intelligent image analysis and interprettation methods. Based on the enumeration of fungal spore classes we have to develop

P. Perner and O. Salvetti (Eds.): MDA 2006/2007, LNAI 4826, pp. 176–182, 2007.

a method that can map this information to the hygiene-relevant parameters. The work we present here reports the results of our study. They show that the proposed methods based on intelligent image analysis and data mining are very suitable to capture the desired information and allow recognizing formerly unknown information that can be helpful to determine the quality of food stuff.

In Section 2 we describe the material used for our study. The image acquisition and sample preparation is explained in Section 3. Section 4 describes the intelligent image analysis and interpretation. The results on the correlation of the recognized fungi spores to the hygiene-relevant parameters are given in Section 5. Finally we summarize our work in Section 6.

2 Material

For the study have been used different quality classes of wheat grains:

1. visual optical perfect grains from a charge where no fungal grains were included,
2. fungal damaged grains,
3. gall-mosquito damaged grains, and
4. visual optical perfect grains taken from a charge of fungal damaged grains.

In total we had 10 samples from each class. Thirty single grains were taken from each sample for further evaluation.

3 Image Data Acquisition

The main problem was to make the coverage on the grains visible under the microscope and make it usable for further digital processing. Therefore we have developed a procedure for taking the coverage from grains and bring it onto a medium that can be placed under a microscope. From there can be acquired a digital image with the help of a digital camera connected with the microscope.

The method of choice was a water-based extraction method. The grains were placed into a boil together with stones. This water-filled boil was shaken for 2 minutes, then the water was filled into a centrifuge and the sediment was put on a slide. This slide was placed under the microscope and a digital image was taken. There are other methods for extracting the coverage from the grain possible but this should not be the main topic of this paper. The resulting digital images are shown in Figure 1a-4a.

4 Intelligent Image Analysis and Interpretation

4.1 Image Analysis

The main aim of the image analysis was to recognize possible fungi spores and process them further for determination of the type of fungi spore. Here we used our novel case-based object recognition method [Perner et al., 2005] developed for recognizing biological objects with high variation. For the architecture of such a system see Figure 5. The case-based object recognition method uses cases that

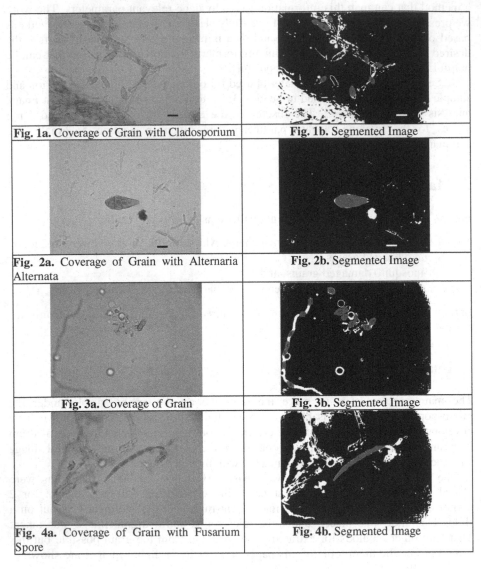

Fig. 1a. Coverage of Grain with Cladosporium | Fig. 1b. Segmented Image

Fig. 2a. Coverage of Grain with Alternaria Alternata | Fig. 2b. Segmented Image

Fig. 3a. Coverage of Grain | Fig. 3b. Segmented Image

Fig. 4a. Coverage of Grain with Fusarium Spore | Fig. 4b. Segmented Image

generalize the original contour of the objects and matches these cases against the contour of the objects in the image. During the match a score is calculated that describes the goodness of the fit between the object and the case. Note the result of this process is not the information about what type of fungi spore is contained in the image. The resulting information tells us only if it is highly likely that the considered object is a fungi spore or not. Further evaluation is necessary to determine the kind of fungi spore. This demonstrates the result in the images, see Figure 1b-4b. One of the main problems of such a case-based object recognition method is to fill up the case base with a sufficient large enough number of cases. We used our procedure described in [Perner et al., 2004] for that. For the study we have 10 different cases, which is not enough as we can see in the image but it allows us to demonstrate the applicability of the method. The method has to be adapted to the specific image quality to show better results as well as more cases have to be learnt by our case acquisition procedure.

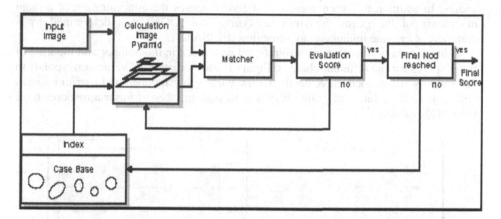

Fig. 5. Architecture of a Case-Based Object Recognition System

4.2 Image Interpretation and Data Mining

After the methods have recognized potential objects that are likely to be fungi spores we have to extract more features from the objects that distinguish the object from the background and different fungi spores. Of course one feature is already the shape information used in the matching process but that is not enough for more detailed recognition. The features that have to be calculated for this kind of objects are the inner structure, texture and gray level information. We haven't done that for this kind of objects considered in this publication yet. But we know from our past research on airborne fungi that it is possible to find automatic extractable features to describe fungi spores and use them for classification into different kinds of fungi spores. It is left to future work to find the right features for the considered fungi spores in this application and to build the feature extraction procedure for them. Based on this feature set we can construct the classifier. We use decision tree induction based on our tool Decision Master [Perner, 2003]. This gives us a good classifier. As the result we will get the information about the kind of fungi spores contained in the image and the number of fungi spores versa the kind of fungi spores.

5 Mapping of Image Information to Hygiene Relevant Parameter with Data Mining

In this study the kind and the number of fungi spores was determined manually since it was a case study and we haven't developed the fully automatic system yet. The aim of the study was to figure out if the proposed methods can bring out information about hygiene-relevant parameters and besides that new information that can be used to control the quality of food stuff. From the 4x10 different samples a data base was created where the columns of each entry show the class, that is the optical visual inspection label, the number of Fusarium spores, the number of Alternaria/Ulocladium, the number of Aspergillus/Penicillium, the number of Cladosporium, the number of fungi spores with unknown classification and the total number of fungi spores. In addition to the enumeration of fungal spores the concentration of a main mycotoxin of the genus Fusarium deoxynivalenol (DON) was determined by a commercial enzyme immunoassay screening (ELISA test).

Table 1-4 shows that there is a significant difference in the number and the kind of fungi spores for the different charges. Figure 6 shows that DON value corresponds to the visually determined class labels. Grain with a low number of Fusarium spores have low DON values and grain charges with high number of Fusariam spores have high DON values.

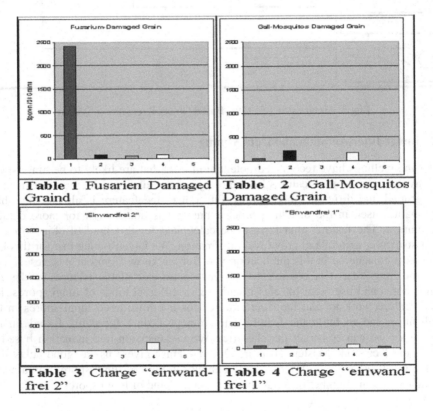

Table 1 Fusarien Damaged Graind

Table 2 Gall-Mosquitos Damaged Grain

Table 3 Charge "einwandfrei 2"

Table 4 Charge "einwandfrei 1"

Decision tree induction with Decision Master [Perner, 2003] on an entropy-based criterion was performed in order to find out the relation between the coverage of fungi spores and the class label (mycotoxin value). The induction experiment shows that there is a relation between the number of Cladosporium spores and Fusarium spores respective the class, see Figure 7. It says that grain charges with a high number of Cladosporium spores will have a low number of Fusarium spores. That means these charges are either perfect charges or gall-mosquitoes damaged charges. Whereas charges with low Cladosporium spores can be either samples with a high number of

Fusarium spores or a low number of Fusarium spores. Note that charge "einwandfrei 2" (visual perfect grains) has been taken out from a sample with Fusarium damaged grains. It seems that the number of Cladosporium spores indicates this fact. The number of Alternaria and Aspergillus spores did not have a significant influence in this experiment.

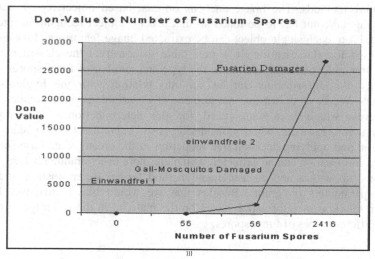

Fig. 6. Don Value to Number of Fusarium Spores

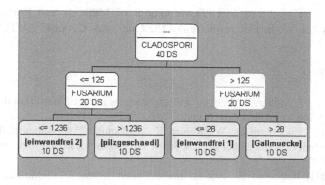

Fig. 7. Decision Tree for the Determination of Grain Quality based on Number and Type of Fusarium Spores

6 Conclusions

We have presented our first results for the detection of hygiene-relevant parameters from cereal grains based on intelligent image acquisition and interpretation methods as well as data mining method. The work is part of a larger project aiming to develop an automatic system to the determination of the quality of cereals in particular wheat. The methods developed so far are protected by industrial property rights [Perner, 2004 and 2005].

We have shown that data acquisition is an important task and that it has to do with more than data base construction as it is in many data mining experiments. The image acquisition method we have demonstrated in this paper works well and can be fully automated. It can also be constructed in such a way that the coverage from each single grain can be taken off and evaluated based on the intelligent image interpretation and data mining methods. The image analysis on case-based object recognition works well for this task but has to be tuned so that a better object recognition rate can be achieved. From each single object can be extracted image features and these features can be used for classification. It is preferable to construct the classifier based on decision tree induction methods. Once the type and number of fungi spores has been determined this information can be set into relation with the hygiene-relevant parameters. We have shown that the number of Fusarium spores correlates with the DON levels which is a value used for the determination of the mycotoxin concentration. However when considering this experiment as a data mining experiment and applying decision tree induction to the created data base some other important information can be extracted which are more or less hidden before. The aim is to come up with a new measurement method for the determination of hygiene-relevant parameters on grains. Besides that we would like to discover formerly unknown relations or information based on the material in the coverage of the grain such as different types of fungi spores.

Acknowledgement

The work is part of the project "Development of Methods and Procedures for an Automatic System to the Determination of the Quality of Cereals" AUTOBONI.

References

[Müller et al., 1997] Müller, H.M., et al.: Fusarium toxins in wheat harvested during six years in an area of Southwest Germany. Natural Toxins 5, 25–30 (1997)

[Herrman et al,, 1998] Hermann, W., Kübler, E., Aufhammer, W.: Ährenbefall mit Fusarien und Toxingehalt im Korngut bei verschiedenen Wintergetreidearten. Pflanzenwirtschaften 2, 97–107 (1998)

[Rodeman, 2003] Rodeman, B.: Auf resistente Sorten setzen. DLG Mitteilungen 3, 44–46 (2003)

[Perner et al., 2005] Perner, P., Jähnichen, S., Perner, H.: Case-Based Object Recognition for Airborne Fungi Recognition. Intern. Journal on Artificial Intelligence in Medicine (to appear, 2005)

[Perner, 2003] Perner, P.: Data Mining on Multimedia Data. Springer, Heidelberg (2003)

[Perner et al., 2004] Perner, P., Jähnichen, S.: Case Acquisition and Case Mining for Case-Based Object Recognition. In: Funk, P., González Calero, P.A. (eds.) ECCBR 2004. LNCS (LNAI), vol. 3155, pp. 616–629. Springer, Heidelberg (2004)

[Perner, 2004] Perner, P.: Procedures and equipment to the automatic and quantitative capture of the portion of seeds or cereals of certain quality, DE 10, 063 769.5 (2004)

[Perner, 2005] Perner, P.: Procedures and equipment to the recognition and classification of fungi spores in cereals, DE 10, 034 504.2 (2005)

Author Index

Lecture Notes in Artificial Intelligence (LNAI)

Vol. 4648: F. Almeida e Costa, L.M. Rocha, E. Costa, I. Harvey, A. Coutinho (Eds.), Advances in Artificial Life. XVIII, 1215 pages. 2007.

Vol. 4635: B. Kokinov, D.C. Richardson, T.R. Roth-Berghofer, L. Vieu (Eds.), Modeling and Using Context. XIV, 574 pages. 2007.

Vol. 4632: R. Alhajj, H. Gao, X. Li, J. Li, O.R. Zaïane (Eds.), Advanced Data Mining and Applications. XV, 634 pages. 2007.

Vol. 4629: V. Matoušek, P. Mautner (Eds.), Text, Speech and Dialogue. XVII, 663 pages. 2007.

Vol. 4626: R.O. Weber, M.M. Richter (Eds.), Case-Based Reasoning Research and Development. XIII, 534 pages. 2007.

Vol. 4617: V. Torra, Y. Narukawa, Y. Yoshida (Eds.), Modeling Decisions for Artificial Intelligence. XII, 502 pages. 2007.

Vol. 4612: I. Miguel, W. Ruml (Eds.), Abstraction, Reformulation, and Approximation. XI, 418 pages. 2007.

Vol. 4604: U. Priss, S. Polovina, R. Hill (Eds.), Conceptual Structures: Knowledge Architectures for Smart Applications. XII, 514 pages. 2007.

Vol. 4603: F. Pfenning (Ed.), Automated Deduction – CADE-21. XII, 522 pages. 2007.

Vol. 4597: P. Perner (Ed.), Advances in Data Mining. XI, 353 pages. 2007.

Vol. 4594: R. Bellazzi, A. Abu-Hanna, J. Hunter (Eds.), Artificial Intelligence in Medicine. XVI, 509 pages. 2007.

Vol. 4585: M. Kryszkiewicz, J.F. Peters, H. Rybinski, A. Skowron (Eds.), Rough Sets and Intelligent Systems Paradigms. XIX, 836 pages. 2007.

Vol. 4578: F. Masulli, S. Mitra, G. Pasi (Eds.), Applications of Fuzzy Sets Theory. XVIII, 693 pages. 2007.

Vol. 4573: M. Kauers, M. Kerber, R. Miner, W. Windsteiger (Eds.), Towards Mechanized Mathematical Assistants. XIII, 407 pages. 2007.

Vol. 4571: P. Perner (Ed.), Machine Learning and Data Mining in Pattern Recognition. XIV, 913 pages. 2007.

Vol. 4570: H.G. Okuno, M. Ali (Eds.), New Trends in Applied Artificial Intelligence. XXI, 1194 pages. 2007.

Vol. 4565: D.D. Schmorrow, L.M. Reeves (Eds.), Foundations of Augmented Cognition. XIX, 450 pages. 2007.

Vol. 4562: D. Harris (Ed.), Engineering Psychology and Cognitive Ergonomics. XXIII, 879 pages. 2007.

Vol. 4548: N. Olivetti (Ed.), Automated Reasoning with Analytic Tableaux and Related Methods. X, 245 pages. 2007.

Vol. 4539: N.H. Bshouty, C. Gentile (Eds.), Learning Theory. XII, 634 pages. 2007.

Vol. 4529: P. Melin, O. Castillo, L.T. Aguilar, J. Kacprzyk, W. Pedrycz (Eds.), Foundations of Fuzzy Logic and Soft Computing. XIX, 830 pages. 2007.

Vol. 4520: M.V. Butz, O. Sigaud, G. Pezzulo, G. Baldassarre (Eds.), Anticipatory Behavior in Adaptive Learning Systems. X, 379 pages. 2007.

Vol. 4511: C. Conati, K. McCoy, G. Paliouras (Eds.), User Modeling 2007. XVI, 487 pages. 2007.

Vol. 4509: Z. Kobti, D. Wu (Eds.), Advances in Artificial Intelligence. XII, 552 pages. 2007.

Vol. 4496: N.T. Nguyen, A. Grzech, R.J. Howlett, L.C. Jain (Eds.), Agent and Multi-Agent Systems: Technologies and Applications. XXI, 1046 pages. 2007.

Vol. 4483: C. Baral, G. Brewka, J. Schlipf (Eds.), Logic Programming and Nonmonotonic Reasoning. IX, 327 pages. 2007.

Vol. 4482: A. An, J. Stefanowski, S. Ramanna, C.J. Butz, W. Pedrycz, G. Wang (Eds.), Rough Sets, Fuzzy Sets, Data Mining and Granular Computing. XIV, 585 pages. 2007.

Vol. 4481: J. Yao, P. Lingras, W.-Z. Wu, M. Szczuka, N.J. Cercone, D. Ślęzak (Eds.), Rough Sets and Knowledge Technology. XIV, 576 pages. 2007.

Vol. 4476: V. Gorodetsky, C. Zhang, V.A. Skormin, L. Cao (Eds.), Autonomous Intelligent Systems: Multi-Agents and Data Mining. XIII, 323 pages. 2007.

Vol. 4460: S. Aguzzoli, A. Ciabattoni, B. Gerla, C. Manara, V. Marra (Eds.), Algebraic and Proof-theoretic Aspects of Non-classical Logics. VIII, 309 pages. 2007.

Vol. 4457: G.M.P. O'Hare, A. Ricci, M.J. O'Grady, O. Dikenelli (Eds.), Engineering Societies in the Agents World VII. XI, 401 pages. 2007.

Vol. 4456: Y. Wang, Y.-m. Cheung, H. Liu (Eds.), Computational Intelligence and Security. XXIII, 1118 pages. 2007.

Vol. 4455: S. Muggleton, R. Otero, A. Tamaddoni-Nezhad (Eds.), Inductive Logic Programming. XII, 456 pages. 2007.

Vol. 4452: M. Fasli, O. Shehory (Eds.), Agent-Mediated Electronic Commerce. VIII, 249 pages. 2007.

Vol. 4451: T.S. Huang, A. Nijholt, M. Pantic, A. Pentland (Eds.), Artifical Intelligence for Human Computing. XVI, 359 pages. 2007.

Vol. 4442: L. Antunes, K. Takadama (Eds.), Multi-Agent-Based Simulation VII. X, 189 pages. 2007.

Vol. 4441: C. Müller (Ed.), Speaker Classification II. X, 309 pages. 2007.

Vol. 4438: L. Maicher, A. Sigel, L.M. Garshol (Eds.), Leveraging the Semantics of Topic Maps. X, 257 pages. 2007.

Vol. 4434: G. Lakemeyer, E. Sklar, D.G. Sorrenti, T. Takahashi (Eds.), RoboCup 2006: Robot Soccer World Cup X. XIII, 566 pages. 2007.

Vol. 4429: R. Lu, J.H. Siekmann, C. Ullrich (Eds.), Cognitive Systems. X, 161 pages. 2007.

Vol. 4428: S. Edelkamp, A. Lomuscio (Eds.), Model Checking and Artificial Intelligence. IX, 185 pages. 2007.

Vol. 4426: Z.-H. Zhou, H. Li, Q. Yang (Eds.), Advances in Knowledge Discovery and Data Mining. XXV, 1161 pages. 2007.

Vol. 4411: R.H. Bordini, M. Dastani, J. Dix, A.E.F. Seghrouchni (Eds.), Programming Multi-Agent Systems. XIV, 249 pages. 2007.

Vol. 4410: A. Branco (Ed.), Anaphora: Analysis, Algorithms and Applications. X, 191 pages. 2007.